IEE COMPUTING SERIES 9

Series Editors: Dr. B. Carré, Prof. S. L. Hurst, Dr. D. A. H. Jacobs,
M. Sage, Prof. I. Sommerville

Computer aided tools for VLSI system design

Other volumes in this series

Volume 1 Semi-custom IC design and VLSI
P. J. Hicks (Editor)
Volume 2 Software engineering for microprocessor systems
P. G. Depledge (Editor)
Volume 3 Systems on silicon
P. B. Denyer (Editor)
Volume 4 Distributed computing systems programme
D. Duce (Editor)
Volume 5 Integrated project support environments
J. A. McDermid (Editor)
Volume 6 Software engineering '86
D. J. Barnes and P. J. Brown (Editors)
Volume 7 Software engineering environments
I. Sommerville (Editor)
Volume 8 Software engineering—the decade
of change
D. Ince (Editor)

Computer aided tools for VLSI system design

**Edited by
G. Russell**

Peter Peregrinus Ltd. on behalf of the Institution of Electrical Engineers

RECID = 10404 - 1

Published by: Peter Peregrinus Ltd., London, United Kingdom
© 1987: **Peter Peregrinus Ltd**.

British Library Cataloguing in Publication Data

Computer aided tools for VLSI system design.
    —(IEE computing series; 9)
    1. Integrated circuits—Very large scale
    integration—Data processing
    I. Russell, G. II. Series
    621.381'73   TK7874

    ISBN 0-86341-093-6

Printed in England by Billing and Sons Ltd.

Contents

List of contributors		ix
Preface		xi
1	**Introduction**	1
2	**Computer aids for VLSI system design – an overview**	9
	2.1 Introduction	9
	2.2 Why VLSI needs CAD	10
	2.3 Basic CAD concepts	11
	2.4 DA tools	15
	2.5 Current and future DA topics	16
	2.6 Conclusion	18
3	**Techniques for circuit simulation**	19
	3.1 Introduction	19
	3.2 Mathematical model and equation formulation	19
	3.3 Solution of nonlinear differential equations	20
	3.4 Standard solution of linear equations	24
	3.5 Relaxation methods	25
	3.6 Conclusions	30
4	**Logic simulation algorithms and techniques**	32
	4.1 Introduction	32
	4.2 Simulation process	33
	4.3 Behavioural modelling of gate level functions	40
	4.4 Conclusions	42

5   **Testing VLSI circuits**   44
    5.1   Introduction   44
    5.2   Design for testability   46
    5.3   Fault modelling   50
    5.4   Practical CAT tools   52
    5.5   Automatic test equipment   63
    5.6   Conclusions   63

6   **Autolayout**   66
    6.1   Introduction   66
    6.2   Placement methods   68
    6.3   Routing   71
    6.4   Conclusions   79

7   **Symbolic design of VLSI circuits**   81
    7.1   Definition of symbolic design   81
    7.2   Short history of symbolic design   82
    7.3   Objectives of symbolic design   85
    7.4   Floor planning and high level design   87
    7.5   Symbolic cell design   87
    7.6   Algorithms for spacing   88
    7.7   Chip assembly   90
    7.8   Algorithms for abutment   92
    7.9   Conclusions   93

8   **IC layout verification**   95
    8.1   Introduction   95
    8.2   Automated design techniques   95
    8.3   Analysis tools for semi-custom design   96
    8.4   Analysis tools for full-custom design   97
    8.5   Exploiting hierarchy   102
    8.6   Databases for IC design   104
    8.7   Conclusions   105

9   **High level languages in design**   106
    9.1   Introduction   106
    9.2   First developments – bristle blocks   108
    9.3   Influence of programming languages   110
    9.4   Special purpose design languages   119
    9.5   Conclusions   122

10   **PLA design tools**   124
    10.1   Introduction   124
    10.2   Structural variants   125

10.3 Computer tools 128
10.4 PLA testability 132
10.5 Conclusions 134

11 Silicon compilation – design and synthesis beyond CAD 135
11.1 Introduction 135
11.2 What is a silicon compiler 136
11.3 Silicon and software compilers 136
11.4 Design leverage and complexity pyramid 138
11.5 Restrictive practices 139
11.6 Review of the field 139
11.7 The Carver connection 144
11.8 Modern approaches 145
11.9 Conclusions 146

12 CAD systems 148
12.1 Systems 148
12.2 Factors common to all design activity 148
12.3 Evolution of IC design systems 150
12.4 Current requirements for IC design systems 151
12.5 The database management system as a solution 151
12.6 Features of database management systems 152
12.7 Differences between IC and general database management
     systems 153
12.8 Nature of IC data 153
12.9 Preparing a data model 155
12.10 Programs interfacing to the IC database 159
12.11 Conclusions 159

13 Data management 160
13.1 Introduction 160
13.2 Requirements 160
13.3 Alternative approaches 161
13.4 Design control in DA-X 163
13.5 Conclusions 166

14 Verification of digital systems 167
14.1 Introduction 167
14.2 What is verification? 167
14.3 LSM 168
14.4 HOL 170
14.5 LTS 173
14.6 Conclusions 176

15 Hardware CAD tools   178
    15.1 Introduction   178
    15.2 Acceleration of logic simulation   179
    15.3 Bottlenecks   182
    15.4 General purpose accelerators   184
    15.5 Selection of accelerators   185
    15.6 Conclusions   186

Index   188

List of Contributors

Chapters 1, 4 and 5
G. Russell, University of Newcastle upon Tyne, UK.

Chapter 2
H. G. Adshead, ICL, Manchester, UK.

Chapter 3
T. J. Kazmierski, University of Southampton, UK (on leave from Technical University of Warsaw, Poland).

Chapter 5
K. Baker, Hirst Research Centre, Middlesex, UK.

Chapter 6
D. J. Kinniment, University of Newcastle upon Tyne, UK.

Chapter 7
P. Ivey, BTRL, Ipswich, UK.

Chapter 8
R. A. Cottrell, UMIST, Manchester, UK.

Chapter 9
M. R. McLauchlan, University of Newcastle upon Tyne, UK.

Chapter 10
E. G. Chester, University of Newcastle upon Tyne, UK.

Chapter 11
S. G. Smith, University of Edinburgh, UK.

Chapter 12
J. D. Wilcock, Plessey Research (Caswell) Ltd, UK.

Chapter 13
D. Warburton, ICL, Manchester, UK.

Chapter 14
M. H. Gill, STC Technology, Harlow, UK.

Chapter 15
A. P. Ambler, Brunel University, Middlesex, UK.

Preface

Over the past decade the use of Computer Aided Design (CAD) Tools in the design of integrated circuits has become well established. However, in order to gain acceptance, in the design community at large, these tools had to be introduced gradually. After a short period of time it was realised that these tools were becoming inadequate as circuit complexities increased due to the advancements made in fabrication technology. Subsequently, CAD research activities were focused on attempting to develop more efficient algorithms to cope with the problem of complexity. This approach, however, could only provide a temporary solution, since the fundamental problem was not that the CAD tools were inefficient, necessarily, but the design style into which they were introduced was simply incapable of supporting the design of complex circuits.

Recently, design styles based on hierarchical decomposition or 'divide and conquer' techniques have evolved which are more amenable to CAD techniques, since they limit the amount of data and design constraints to be processed at each stage of the design cycle, resulting in reduced design times and costs through the more efficient use of man/machine resources.

Although, CAD tools have now been integrated into each stage of the design process, the IC design community, which is continually increasing, is not fully aware of the arsenal of CAD tools available for integrated circuit design. Consequently the objectives of the 1st IEE Vacation School on CAD Tools for VLSI System Design, held at the University of Newcastle upon Tyne in July 1985, were twofold. First, to review the range of CAD Tools available for design and, thereafter, in the light of projected increases in chip complexity, anticipated through improvements in device technology, to discuss the limitations of these tools. Second, to introduce the next generation of design tools and associated design methods which should restrict, to within acceptable limits, the major obstacles to the pervasive use of VLSI, namely design times and costs. The contents of this book comprise the lectures given during the course, which are self contained and cover a wide range of topics related to the use of CAD tools in the design of integrated circuits.

G. Russell

Chapter 1

Introduction

What is CAD? CAD, in the context of the design of integrated circuits can be defined as the use of computers for the collection, manipulation and storage of all the data associated with the design. More specifically, CAD tools are used to evaluate design alternatives and tradeoffs, to construct the basic components used in the design, to assemble and interconnect these components into the final layout of the design and finally to generate the data and test files used in mask making and testing the device after fabrication.

It is well established that CAD techniques are essential to the design of Very Large Scale Integrated (VLSI) Circuits in order to process, efficiently and cost effectively, the vast amount of data associated with the design of these complex devices. The reason given for using CAD in VLSI circuit design, to some extent, belittles the important role that CAD now plays in ensuring that IC products from a given company retain the competitive edge in the market place. The factors which make an integrated circuit competitive are performance, functionality, cost and turn around time. In general performance is a function of the processing technology, this has reached a very advanced state of development, to the extent that further gains in performance through improvements in technology are extremely expensive. Functionality reflects the ingenuity of the designer, again good designers are expensive; the functional capability of a circuit can be duplicated and superseded, hence the competitive edge gained in this way is marginal. The remaining factors are cost and turn around time, these are directly associated with CAD systems; hence the efficiency of the CAD systems used in IC design is now playing an extremely important role in maintaining a competitive edge in the market place for IC products. In order to maintain a competitive edge it is essential that the CAD system has a wide range of CAD tools, the ability to support various design styles and have an extensive library of functional cells. If the CAD system is not to fade into obsolescence it must also contain a flexible database management system, since this is the foundation on which all efficient CAD systems are built. The database system permits new tools to be interfaced, readily to the system, it controls the versions of cell libraries used in a design, it ensures the integrity of design changes throughout the various representations of a design and also allows changes in design style or technology to be accommodated, readily, in the system. A further factor which affects the efficiency of the CAD system is the types of tools that it contains. If the system has many analysis tools, it infers that the design time will be long and expensive, since the CAD system will impose few constraints on the designer, requiring long verification runs to ensure that the layouts are error free. Although the efficiency of these systems can be improved by developing analysis algorithms which exploit the structure of the layout in the verification procedures or have the analysis algorithms mapped onto special purpose computer architectures, the competitive edge gained in this way only lasts a short time. However, if the system contains synthesis tools, design times are drastically reduced since the designer is constrained; the essence of these tools is 'correctness by construction'. Furthermore, the competitive edge gained will last longer since the capabilities of these systems are not fully explored, leaving ample opportunity for improvement and also the capabilities of synthesis systems are more proprietary hence more difficult to duplicate.

Although the concept of using CAD tools in the design of digital systems has been around since the mid 1950's, it has only been used in IC design since the mid 1960's. The reason for this is twofold; first the concept of using computers in the design of digital systems was brought into disrepute, throughout the late 50's and early 60's due to the inadequacy of the computer systems (in terms of computational power, storage facilities and availability) to solve the problems which occurred in the design of digital systems. Second, the complexity of the task in designing integrated circuits at that time, was still amenable to manual methods, and there was a reluctance to change to CAD techniques which, in the past, had been shown to be inadequate. However, as circuit complexity increased manual methods were becoming inadequate and alternative design techniques were sought. By this time there had been a considerable improvement in computer technology, with respect to both hardware and software. Subsequently, some simple layout tools evolved which permitted the designer to digitise a layout into a computer and have a plot produced to determine what errors had been made, and thereafter have the drive tapes for the mask making equipment generated automatically. The introduction of interactive graphics greatly enhanced the designers capability to edit circuit layouts stored in the computer. At the same time designers were becoming interested in the use of circuit simulation as a means of verifying the operational characteristics of their circuits. Previously breadboarding techniques had been employed to simulate the behaviour of the circuit, however due to the differences in device size and parasitic capacitances, particularly as IC's became more complex, breadboarding could no longer be used as a means of checking, satisfactorily the performance of the circuit. As the complexity of integrated circuits increased, simple functions which were interconnected on a printed circuit board were now integrated into the same circuit. Consequently

IC designers had a requirement for placement and routing programs, which they subsequently borrowed from the printed circuit board designers and modified to suit their own requirements. Also, the complexity of the integrated circuits being designed was becoming too large to be efficiently simulated at circuit level, consequently gate level simulators were also borrowed from the designers implementing systems on printed circuit boards. However, by this time some designers had become interested in developing their own suites of CAD tools and several tools evolved which were directed at the time consuming tasks, for example, design rule checking, connectivity checking, layout to function verification. The increased functionality of integrated circuits, also, permitted the development of computer systems with better performance characteristics, which allowed computational intensive CAD tools for example, fault simulators and automatic test pattern generators to be developed. A major factor which influenced the acceptance of CAD tools into the IC design process was that they were introduced incrementally and were oriented towards a psuedo-manual design style. In this design style the layout is considered to be a monolithic entity without any structure, consequently as circuit complexity increased the capabilities of the CAD tools were far exceeded due to the vast amounts of data to be processed. To combat the complexity problem a new generation of CAD tools evolved, which could be classed as revolutionary, since they imposed a design style upon the user. It constrained the designer to using function blocks, for example ROMs, RAMs, PLAs and Shift Registers, whose regularity could be used to advantage in improving the efficiency of the CAD tools which would process the layouts. Simple design languages also evolved to describe the layout of these regular structures at a high level, from which a detailed layout description could be generated automatically to prescribed design rules. PLA layouts could be derived directly from their logical specification. This style of design required the designer to consider the circuit to comprise functions more complex than basic gates, this resulted in the development of high-level simulation tools, which also improved the simulation efficiency of large circuits. The basic trend in the CAD tools, currently under development, is that they perform more of a synthesis rather than an analysis function, for example PLA generators and more recently silicon compilers, whose objective is to automatically generate the complete layout of a circuit from its behavioural description. The trend towards the development of synthesis emphasises the importance of these tools in maintaining the competitive edge in a CAD system.

The individual chapters in this book describe some of the tools discussed in the preceding paragraphs in more detail.

Chapter 2 discusses, in general terms, the need to use computers in the design of VLSI circuits. The VLSI design process is described as a major data processing problem involving vast amount of data, complex algorithms, CPU intensive processes and many aspects of man machine interaction. Typical CAD tasks are outlined and the distinction is drawn between Design

Automation and Computer Aided Design. The Chapter ends with a brief survey of future trends in CAD tools for VLSI design.

Chapter 3 is the first of three chapters discussing tools which may be classified as behavioural analysis tools for both faulty and fault free circuits. This chapter is concerned with techniques of circuit simulation and describes the mathematical modelling and equation formulation of non-linear circuits with lumped elements. Techniques used to provide approximate solutions to systems of non-linear differential equations are also discussed and compared with respect to their numerical efficiency and stability properties. The transformation of a system of non-linear ordinary differential equations into non-linear algebraic equations and a final to a set of linear algebraic equations is also described and a simplified algorithm, which may be used in a classical circuit level simulator for this purpose, is outlined. Finally, in view of the increase in the size of circuit to which this level of simulation is applied and the subsequent increase in solution time, Relaxation Methods are introduced as a means of reducing equation solution time.

Continuing with the topic of simulation Chapter 4 describes the 'anatomy' of a gate level simulator. The basic components which make up a simulator are described together with two techniques for modelling a circuit for the purpose of simulation, namely the Compiled Code and Table Driven Model. The basic simulation algorithm is outlined, together with the basic techniques for evaluating the change in logic state on the output of gate and scheduling events or gate changes in the simulator. The factors which affect the accuracy of the simulation results are also discussed. The problems of using standard simulators to model devices, for example pass transistors, which can have a dynamic state or exhibit a bilateral switching characteristic are outlined. The chapter is concluded with a section outlining the advantages of describing functions using a Behavioural Modelling language and the possibility of using this language to describe analogue functions so that they can be simulated in a digital environment.

A major issue in the design of VLSI circuits is that of testing the devices after fabrication, this topic is addressed in Chapter 5. The basic problems in testing VLSI devices are discussed and this is followed by a brief description of the various Design for Testability (DFT) techniques, namely Ad-hoc, Classical Structured and Neo-Structured methods. Although DFT facilitates the testing of circuits, the problem, in general of generating the test patterns in the first instance, still remains. An important aspect of test pattern generation is fault modelling, the major categories of faults considered in testing are outlined, namely stuck-at-faults, bridging faults, stuck-open faults and pattern sensitive faults. An essential adjunct to test pattern generation is Fault Simulation which is used to determine the fault coverage of the test patterns, the techniques used, namely Parallel, Deductive and Concurrent Simulation are outlined. The basic technique for generating tests in combinational circuits using the Boolean Difference Method and the D-Algorithm are also described, together with a technique called PODEM which was designed to generate tests in very large combinational circuits. The chapter is concluded with a brief description of Testability Analysis Tools and some comments on Automatic Test Equipment.

Chapter 6 is the first of six chapters which deals with topics related to the physical synthesis of a layout. As the complexity of an integrated circuit increased the functions normally connected together on a printed circuit board were now integrated at chip level and designers had a requirement for placement and routing tools, these topics are discussed in Chapter 6. In order to make the layout and interconnections of modules on either an integrated circuit or printed circuit board a tractable problem, the processes of placement and routing are considered separately. The placement algorithms which are outlined in this chapter are the Cluster Development and Force Directed techniques, techniques to improve placements are also discussed together with methods of avoiding wire congestion during the routing phase. An essential part of the layout problem is that of routing the interconnections between the placed modules, the basic method described, to determine the interconnection path, is Lee's Algorithm, together with several heuristics to reduce the routing time. The chapter concludes with a description of Channel and Hierarchical routers.

The traditional method of designing basic logic elements in a layout comprised drawing out the individual shapes on each mask layer required to realise a transistor, contact etc., in the physical circuit. This process is very time consuming and error prone. Recently, basic cell design has been made more efficient by the use of symbolic design tools, these are discussed in Chapter 7. The major techniques described are the Fixed and Relative grid approaches, in the Fixed Grid technique all the design rules are obeyed on entry of the cell description, however using the Relative Grid approach it is necessary to modify the layout to obey various design rules, hence a necessary adjunct to this approach is a Compactor program; several compaction algorithms are briefly described. The technique of using symbolic methods at a higher constructional level are discussed in the form of Floor Planning. The chapter is concluded by a brief description of chip assembly techniques and cell abutment algorithms used in the formation of complex functional blocks from subcells.

Chapter 8 discusses the range of layout analysis tools used in IC design to detect any physical design errors which will either reduce the yield on the fabricated circuit or realise a logic function other than that intended by the designer. The layout tools are classified as either semi-custom or custom, since the semi-custom design style obviates the necessity of using certain analysis tools required in custom design. The major analysis tools used in semi-custom design are wire delay extractors required to determine circuit delay for post layout simulation and circuit verifiers which check that the interconnection of modules on the layout matches that described at a higher level of abstraction.

In a full custom design the designer has fewer constraints, and hence the layout is more likely to contain errors, consequently a wider range of tools are available, for example layout rule checkers, circuit extractors which perform layout to circuit verification, parameter extractors for circuit simulation, electrical rule checkers and netlist to layout verifiers. Techniques to improve the efficiency of these tools by exploiting hierarchy in the circuit are also discussed together with the problems encountered in designs which have overlapping cells. The difficulties in checking the correspondence between the actual and extracted circuits are discussed, together with the role that a good database plays in this function.

Chapter 9 describes the use of High Level languages in the design of integrated circuit layouts. The chapter starts with a brief review of the manual techniques of designing IC layouts and the description of some low level mask layout description languages, outlining their limitations. Some of the unique aspects to be considered when attempting to define a high level language for IC design are identified together with advantages to be gained by using high level languages in the design process. The Bristle Block system is then identified as an example of a first attempt at using high level languages in the design process. The basic disadvantages of the Bristle Block approach are subsequently outlined, and the use of procedural design languages, in overcoming these disadvantages, are discussed. The issues involved in attempting to capture structural and behavioural aspects of a design using procedural types of design languages are also discussed and several examples of existing languages are given. The chapter concludes with a section on special purpose high level languages for IC design in contrast to the use of standard languages with embedded procedures.

A major problem in the design of datapath circuits is the implementation of the control logic for the datapath. The most effective way of implementing this logic is to use a PLA design style. This task has been made easier by the introduction of PLA generators, which may be described as a crude form of silicon compiler and are discussed in Chapter 10. The standard input to a PLA generator is a set of Boolean equations and the output is an regular array structure which realises the Boolean equations. A straight implementation of the Boolean equations, however, results in an inefficient structure in terms of size and performance. Subsequently optimisation techniques, for example logic minimisation and PLA folding methods, to improve the performance of the PLA are discussed. Variations of the basic PLA structure are also outlined, for example Weinberger Arrays and Storage Logic Arrays together with the formation of FSMs from PLAs by including latches between the secondary input and outputs. The chapter concludes with a section dealing with the testing of PLAs and methods used to enhance their testability.

Chapter 11 is the last of the six chapters considering the different aspects of the synthesis of layouts and is concerned with a description of Silicon Compilers. The introduction of the concept of silicon compilers into the design process is considered revolutionary as it removed the need to know about the details of the low level implementation of a design, permitting the designer to work at the higher levels of architectural or functional abstraction. The chapter starts with the basic definition of silicon compilation and the parallels between silicon and software compilation are discussed. Thereafter several Silicon Compilers are described briefly, for example FIRST, Chipsmith and Metasyn. The chapter concludes with a discussion on the limitations of current silicon compilers in terms, for example, of the input description to the compiler, the amount of feedback or interaction with the designer and their general capabilities.

Chapters 12 and 13 are concerned with the requirement for database management in CAD systems for IC design. Chapter 12 starts by defining the factors common to all design activities and subsequently discusses the evolution of IC design systems. If a CAD system is not to become obsolete, it must be flexible with respect to changes in design style and technology and new tools must be integrated easily into the system. Furthermore a CAD system must be capable of controlling design changes and ensuring that any alterations to a circuit are reflected in the various levels of description of the circuit. In view of the diverse requirements of a CAD system, the introduction of a centralised database is proposed as a solution to the problem. The features of a database management system are then discussed, and the characteristics peculiar to a database management system for IC design are outlined. The nature of IC design data is discussed and a data model for use in a database management system is described. The chapter concludes with a brief description of some of the programs which interface with the database during the course of a design.

Chapter 13 starts with the importance of the data management in IC design and briefly outlines the requirements of a database management system including a requirement that it should be independent of the host system and have the ability to function in a networked environment. The different approaches to database management systems are outlined, namely the file based system and the integrated database, the advantages and disadvantages of each are then discussed. Design control in an integrated database is then described with reference to the DA-X system.

The last two chapters in the book are concerned with recent advances in CAD Tools. Chapter 14 discusses the technique of formal verification applied to the design of digital systems. At present, the technique of verifying the function of a digital system is to simulate the design with a given set of input patterns. This form of verification is incomplete since all that it proves is that the circuit functions correctly for the set of applied inputs. However, formal verification techniques are input independent and have been developed to prove the functional equivalence of different representations of a design. Since the concept of formal verification is alien to most designers the chapter starts with a definition of the verification process and outlines the principles involved by using a verification system called LSM on several small example circuits. Having outlined the basic principles involved two other systems are described.

Over the past decade improvements in fabrication techniques have permitted more complex designs to be integrated onto a single chip. This increase in complexity, however, has placed greater demands upon the efficiency of our CAD tools. The efficiency of the tools is limited by the use of general purpose computers as the host machine for the CAD programs. The efficiency barrier, however, can be overcome through the use of hardware accelerators which exploit the parallelism and concurrency existing in most CAD Algorithms. Chapter 15 deals with the development of hardware accelerators. Although the accelerators demonstrate an improved efficiency of several orders of magnitude over the corresponding software tools, they have certain deficiencies and these are described. Thereafter the criteria for selecting an accelerator are outlined. The chapter concludes with a discussion on the factors which may reduce the need for hardware accelerators, for example, a change in design style or the development of more efficient software algorithms, however in this case the algorithms would become even more efficient if they were realised as hardware accelerators.

It should be understood that, although a wide range of topics, related to the application of CAD techniques to the design of integrated circuits, are covered in this book it can only be considered as an introduction to the subject which has expanded, immensely, over the past decade. Additional information on individual aspects of the CAD tools discussed can be obtained from the references cited at the end of each chapter.

Chapter 2

Computer aids for VLSI system design – an overview

H. G. Adshead

2.1 Abstract

This set of notes attempts to provide a fairly high level and general view of how and why computers are used in the process of designing VLSI systems. The nature of the VLSI design problem is discussed together with the nature of the art of CAD. The VLSI design tools themselves are only briefly discussed as these are the subject of other lectures in this series. In conclusion, a brief survey is made of some of the more recent developments and future trends.

2.1.1 Audience

Techniques involved in computer-aided design of VLSI systems form a fascinating blend of hardware and software skills and problems. This paper is addressed both to the newcomers to CAD programming and to VLSI systems designers who are finding increasingly that they need to apply CAD design techniques to their design problems.

2.1.2 Introduction

Electronic systems play an ever-increasing role in our society. This revolution has, of course, been brought about by the increasing use and availability of micro-electronic components. The trend towards ever-increasing levels of integration (VLSI) appears to be never-ending. These VLSI systems and components just cannot be designed without the ever-increasing use of computerised design tools.

2.2 Why VLSI needs CAD

2.2.1 Scope of design tasks

Computer aids are used in a very wide range of problems associated with the design, manufacture, test and maintenance of complex electronic systems.

There are many problems associated with the design of circuit devices, beam tubes, magnetic devices, optimisation, manufacturing processes, tolerancing, noise, thermal and mechanical environment, etc. In most of these areas ad-hoc computer programs are often vitally necessary and very effective. This class of CAD is discussed no further as the intention is to focus on VLSI systems.

A VLSI system is usually intended to exist or interact with a real-world environment and may contain much application software and microcode. In order to ascertain the full specification of the electronic hardware it is usually necessary to use further ad-hoc programs, simulations, performance predictors, system analysers, etc, which need to model both the desired system and the associated environment. This class of CAD is discussed no further as the intention is to focus on the hardware complexity of transistors and interconnect.

The scope of the problem may at first sight seem limited, but in fact represents a major set of data processing challenges and opportunities. A typical VLSI system will consist of many layers of physical interconnection – chips, boards, backplanes, cables, etc, involving the order of millions of wires interconnecting millions of transistors.

2.2.2. VLSI Trends

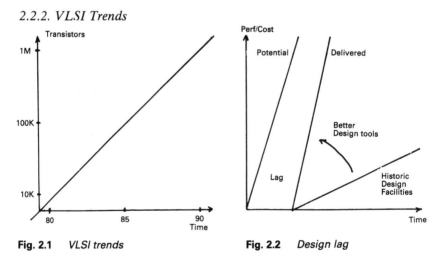

Fig. 2.1   *VLSI trends*   **Fig. 2.2**   *Design lag*

Processing technologies emerging in the period 1985–1990 will allow the manufacture of VLSI devices containing from 100 thousand to 1 million transistors (Fig. 2.1). If system designers do not exploit these devices their products rapidly become obsolete. For any given design method, it is often found that development timescales increase alarmingly as the design size increases. A key role of CAD methods is to reduce the gap between technology potential and delivered technology (Fig. 2.2).

2.2.3 Typical VLSI CAD tasks
Briefly, CAD VLSI tools are employed to:

(1) Store the design data.
(2) Interactively edit or alter the design.
(3) Generate numerical control data to drive manufacturing or test machinery.
(4) Validate the design to remove errors.
(5) Automatically synthesise and contribute design decisions – typical problems involve the generation of test data and the layout of circuits, components, wires etc.
(6) Provide information or extracted views of the design data to assist the human designer or the engineer in the field

2.3 Basic CAD concepts

'Design' can be summarised as 'an iterative decision-making process'. For VLSI, many millions of design decisions have to be made. CAD can help to keep track of these decisions. DA (Design Automation) helps by automatically making many of these decisions.

2.3.1 Design Areas

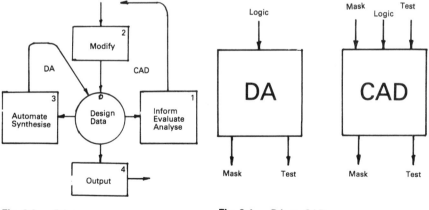

Fig. 2.3   *DA areas*

Fig. 2.4   *DA -v- CAD*

Fig. 2.3 shows the 5 main ways in which computers can be employed to assist the design process. The central theme is the concept of storage of design data (the database). It is important to retain not only the current state of the design but many of the previous releases, versions, etc, so that it will be possible to deduce changes between any two states.

Box 1 summarises a whole class of programs that can inform designers about the current views of the design, or evaluate, simulate or analyse the design to perform design error checks.

Box 2 indicates programs to input design data or more importantly, modify and correct the design. A good CAD system will follow up all ramifications of a small change to a large structure. The feedback of errors to the designers and subsequent editing forms the CAD (computer-aided design) loop.

Box 3 summarises a whole class of programs that can automatically contribute design decisions. These are the design algorithms or synthesis procedures. It is important to observe that this design loop DA is an exact equivalent to the CAD loop where a computer program has replaced the human designer.

Box 4 relates to a class of programs that output design to other processes. It is important that these programs should merely translate the format of the data and not accidentally contribute design decisions. All design decisions should be recorded in the design database to ensure consistency of all design descriptions.

2.3.2 DA or CAD

Fig. 2.4 indicates the key difference between the DA and CAD approaches to a problem. For example, a DA subsystem may take in logic equations and automatically generate the circuit mask details together with test patterns to validate the manufacturing operation. The equivalent CAD subsystem would require manual design of the three sets of data (logic, mask, test) and provide mechanisms for cross-checking that the tests and the mask details agree and line up with the required logic design.

In general, the DA approach will only work within certain constraints, will take longer to program, and will tend to generate less efficient design solutions. The CAD approach is easier to program, is more flexible and allows human designers to create novel or more efficient designs.

However, the CAD approach is much more labour intensive and can be prone to undetected design errors. The DA approach requires more computer power and leads to much safer solutions with high design productivity.

In practice, most design methodologies will require a fair mix of these two approaches. For example, designs may be 95% automated and then finished off with interactive design techniques.

2.3.3 CAD, CAM, CAE terms

The terms CAD, DA, CAE, etc, are used rather imprecisely. There is probably no universally-accepted distinction. However, in general:

DA – Design Automation: implies a higher level of productivity.
CAD – Computer Aided Design: relates to graphical data.

CAE – Computer Aided Engineering: relates to electrical data.
CAM – Computer Aided Manufacture: relates to mechanical data.
CAT – Computer Aided Testing: relates to the test gear.
CIM – Computer Input to Manufacture: implies associated CAD.

2.3.4 Development routes

To manage the complexities of a VLSI system design, it is important to first define for each major component technology the design method, the development route and the associated DA tools and DA framework.

There are many VLSI design styles, eg: parametric cells, gridless symbolic, Weinberger arrays, PLAs, etc. In general, each of these approaches will have different DA tools. The definition of the development route describes the sequence of each design step, the audit points, the DA tools to be used, and validation procedures. Often, special DA programs will be required to interface to silicon suppliers and to link various tools together.

2.3.5 Characteristics of VLSI design problems

VLSI design is a major data processing task involving large quantities of data, complex design algorithms (that may themselves be erroneous), CPU intensive runs and many aspects of man-machine interaction.

2.3.6 VLSI CAD skills

It is sometimes felt that there are no specific 'CAD' skills as the art involves a mix of software engineering, data processing techniques and a particular branch of engineering.

A good DA programmer needs to understand database technology, design algorithms, language techniques, computer graphics, networking, man-machine interaction methods, artificial intelligence techniques, VLSI design methodologies, testing strategies, special DA hardware and VLSI manufacturing and process technologies, and more!

2.3.7 History of VLSI CAD

Early DA techniques evolved in the 1955–70 period within the design groups of mainframe computer manufacturers, mostly concerned with layout wiring and multilayer boards. In the 1970–80 period, semiconductor manufacturers created circuit analysis tools and graphical capture of polygon mask shapes. During this period, simulation techniques became necessary. Much good CAD software was generated in university groups. The period 1980–85 has seen a tremendous growth in new VLSI design methodologies, low cost hardware and individual design workstations and a large number of CAD companies selling design systems. Current VLSI design methods involve a wide mix of system level and circuit level problems.

2.3.8 Nature of VLSI CAD data

Electronic design involves working with DESIGN UNITS, eg: a CPU, a multiplexor, a register or a latch cell. For each unit there can be many VIEWS. Different design operations will need different views of the object (Fig. 2.5) eg: logical behaviour, electrical behaviour, layout shape, heat, power consumption, etc.

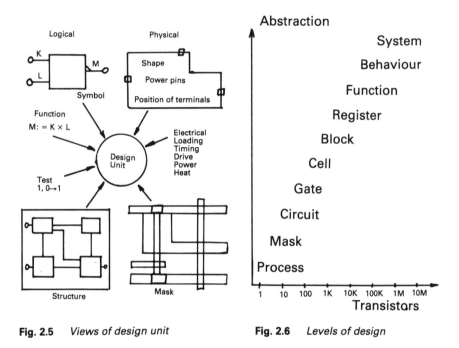

Fig. 2.5   *Views of design unit*   **Fig. 2.6**   *Levels of design*

For each design unit there will be an appropriate LEVEL OF ABSTRACTION (Fig. 2.6). In general, the bigger the unit in terms of numbers of transistors, the more appropriate it will be to use a high level language or summary view of its properties.

2.3.9 Nature of VLSI design tasks

A very simplified summary of the VLSI design process is to create a full hierarchic set of views of the components of the system.

The design tasks then boil down to verifying the correctness of the total set of views. This involves synthesising one view from another (correctness by construction) or cross-checking that all views of a unit are consistent and that each high level view agrees with the more detailed level of design in the level below.

2.3.10 DA systems
Often, the key reason for using a DA system is because of a particular automatic technique. The algorithms get all the attention but are often only a small proportion of the total system. 95% of a DA system will relate to relatively mundane data processing tasks.

2.3.11 DA programming
Historically, Fortran was used for CAD programs with some use of Algol, Algol68 and Fortran4. Recently Pascal has been used in many systems with some use of BCPL. Portability over a wide range of low-cost hardware has caused a de-facto standard in this area of the use of UNIX as the operating system. This, in turn has encouraged the use of C in CAD programs. Current DA research is often focused on AI/Expert system approaches and some interesting use of LISP and Prolog is being made. This is also leading to the use of the LOOPS and SMALLTALK class of programming environment. Modern DA systems pay a lot of attention to longevity, good system interfaces and portability.

2.4 DA tools

As implied above, a comprehensive VLSI design system will contain between 100 and 300 separate tools. This section attempts to briefly summarise and classify the key tools.

One key property of VLSI systems is the fact that a large proportion of the wires (design decisions) are buried inside the VLSI chips. It is usually impossible to modify these wires and often exceedingly difficult to diagnose the source of a design error in the hardware prototypes. This, in turn, puts considerable emphasis on design verification, design state control and precision of development routes and design management.

2.4.1 Database and data management
Vitally important.

2.4.2 Design capture
Much use of personal workstations and local interactive graphical techniques together with local design checks preferably in real time.

2.4.3 High level specification languages
Languages and associated simulators to take the highest possible level of description of each unit. There is growing interest in languages that promote good design, allow communication to other designers, allow efficient simulation and can be used for formal proof techniques.

2.4.4 Simulation tools

Animation modelling of logic is currently the best way of detecting many classes of design error. A model of a unit is constructed at the appropriate level (or multi-levels) and input stimuli or test patterns are applied. Interactive browsers are later used to analyse the results and simulated node behaviour.

2.4.5 Static validation tools

Simulation is a very expensive and imprecise technique. Where possible, static checks should be used to detect certain classes of error. A good example of this is a timing path checker. For synchronous systems, in a single pass through the logic, it is possible to predict the statistical spread and timing behaviour at the input to each clocked latch. A single reverse pass through the logic will diagnose the critical timing paths in the system.

2.4.6 Layout tools

These often depend on particular design methodologies. The main object is to define the placement of each sub-component, the choice of input/output terminals, and the routing of all interconnection nets. Many algorithms exist in this area and solutions are often 100% automatic.

2.4.7 Test generation tools

Many approaches exist in this area which gets increasingly important as the size of chips increases. Automatic test pattern generation techniques and fault grading programs are needed but much interest has focused recently on tools to support BIST (Built-in Self-test).

2.5 Current and future DA topics

This section attempts to present a very quick look at some CAD areas that are currently attracting much research and development thought.

2.5.1 Silicon compilers

This is a vast and fascinating subject. A wide range of techniques are currently being created to generate large structured cells (cell compilers). In this way quite large regions of silicon (10K transistors) can be generated quickly from a very small input statement. In some circumstances these cells can be automatically laid out and interconnected from a small set of specification statements (silicon compilers by commercial standards). The use of AI rule-based approaches has led to some cases of direct chip synthesis from a functional specification (Fig. 2.7).

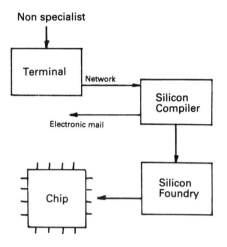

Fig. 2.7   *Silicon compilation*

2.5.2 *Use of AI techniques*
In the areas of test, layout, verification, user interfaces and logic synthesis, LISP and Prolog are being applied to increase design productivity. In many cases this has led to more efficient solutions. As the knowledge base increases and with the increasing use of special AI hardware engines, it is clear that this will be a major growth area.

2.5.3 *Convergence with software engineering*
It is clear that the design (at the systems level) of VLSI systems is fast becoming more and more like the software design problem. This suggests the use of IPSEs (Integrated Project Support Environments).

2.5.4 *Collaborative CAD developments*
There has been much recognition in Japan, Europe, USA and the UK of the strategic importance of VLSI technology and the associated CAD tools and techniques. It is felt that the scale of development required (200–500 man years) is usually too great for any single company and as a result, various funding agencies have promoted several national and international projects, eg: Alvey, Esprit, MITI, VHSIC. One of the most significant projects is MCC in America. Here, a consortium of 12 companies is developing a CAD VLSI system totally in LISP on special AI hardware.

2.5.5 *Special-purpose CAD hardware*
As the scale of a VLSI problem increases in terms of the number of transistors, many of the algorithms (eg: simulation) increase in run time. Often, the CPU needed for this sort of design increases alarmingly according to a power law such as n^2 or n^3. There are now several companies offering accelerator bases

networked to the design capture terminals. These tack simulation, routing, layout verification, placement, fault simulation, etc. Often speed-up factors of 1000 can be achieved in terms of the time to do the job. However, a major overhead in terms of loading and unloading data to and from the special box is found and the actual elapsed time speed-up factor for a specific task is only 10–20 times.

2.5.6 Growth of commercial CAD
Over 100 companies now sell CAD tools worldwide. The implication of this is that most DA systems just have to include some commercial components. The role of a CAD programmer will increasingly involve the creation of data bridges between tools.

2.5.7 Emerging DA standards
To achieve a VLSI system design will increasingly involve data interfaces between companies, suppliers, cell generators, etc. At the world level, EDIF (Electronic Data Interchange Format) is rapidly becoming a standard for interchange of cell data.

2.6 Conclusion

Computer aids for VLSI systems design are increasingly important and cover a fascinating and wide range of techniques. VLSI systems just cannot be designed without a considerable investment in CAD.

Chapter 3

Techniques for circuit simulation

T. J. Kasmierski

University of Southampton, UK
On leave from Technical University of Warsaw, Poland

3.1 Introduction

The time-domain simulation of an electronic circuit modelled at the basic component level is a considerable task involving most advanced numerical and programming methods.

Simulators developed in the early 1970s resolved many of the problems by using implicit integration and sophisticated techiques for solving sparse equation (Chua and Lin (1)). Amongst the most successful are SPICE and ASTAP which are still used extensively in the design and verification of moderately large circuits with up to several hundred nodes. In recent years relaxation based simulators emerged that could speed up the simulation of big circuits by at least two orders of magnitude.

Most advanced hybrid simulators (Newton et al (4), Rabbat et al (6), Lelarasmee et al (8)) can perform exact circuit simulation for highly coupled circuit blocks combined with less accurate timing analysis based on nonlinear Gauss-Seidel-Newton relaxation for loosely coupled blocks. Some circuit parts can be analysed at logic level. Relaxation techniques are specifically emphasised in this contribution due to their growing popularity and the advantages offered to circuit simulator designers. Particular attention is devoted to nodal equations of circuits with grounded capacitors at each node because of their remarkable numerical properties.

3.2 Mathematical model and equation formulation

A nonlinear electronic circuit with lumped elements can be modelled by a set of nonlinear ordinary differential equations of the form

$$f(v(t), \dot{v}(t), t) = 0 \quad t \geq 0 \tag{3.2.1}$$

with the initial value $v(0) = v_0$, where $v(t)\varepsilon R^N$ is the vector of primary circuit variables which can be, in general, node voltages, branch currents, capacitor charges and inductor fluxes, $v(t)\varepsilon R^N$ is the vector of time derivatives of v and f: $R^N \times R^N \times R^1 \rightarrow R^N$ is a continuous function with respect to v, $\dot{v}$ and t.

In particular, if the circuit consists solely of linear or nonlinear voltage controlled capacitors, conductors, current sources and independent voltage sources (dc or time-varying) then nodal formulation can be used and the equation (3.2.1) can be presented in the following form

$$C(v(t), u(t))\, \dot{v}(t) + q(v(t), u(t)) = 0 \tag{3.2.2a}$$

$$v(0) = v_0 \tag{3.2.2b}$$

where $v\varepsilon R^N$ is the vector of unknown node voltages with initial values v_0, $u\varepsilon R^r$ is the vector of all independent voltage sources, q: $R^N \times R^r \rightarrow R^N$ is a continuous function each component of which represents the sum of currents charging the capacitors at each node, C: $R^N \times R^r \rightarrow R^{N \times N}$ is a matrix function continuous with respect to $v(t)$ and $u(t)$.

In addition, if every node has a capacitor, called a grounded capacitor, connected to either ground or the terminal of a voltage source then C is a symmetrical strictly diagonally dominant matrix with positive main diagonal entries where $C_{ii} > 0$, $i = 1, \ldots, N$ is the sum of the capacitances of all capacitors connected to the node i, $C_{ij} \leq 0$, $i \neq j$, $i,j = 1, \ldots, N$ is the negated value of total capacitance between nodes i and j and

$$|C_{ii}| > \sum_{\substack{j=1 \\ j \neq i}}^{N} |C_{i,j}|, \quad i = 1, \ldots, N \tag{3.2.3}$$

Provided there are grounded capacitors at every node, the node voltages establish a valid choice of state variables and the nodal equation (3.2.2) is also a state equation of the circuit. It can be easily transformed into its canonical form

$$\dot{v}(t) = -C^{-1}(v(t), u(t))q(v(t), u(t)) \tag{3.2.4a}$$

$$v(0) = v_0 \tag{3.2.4b}$$

since C is a nonsingular matrix.

3.3 Solution of nonlinear differential equations

The approximate solution of (3.2.1) can be obtained by replacing the time derivatives by a discrete algebraic estimation

$$h_n \dot{v}_n = \sum_{i=0}^{P} \alpha_i v_{n-} \cdot + h_n \sum_{i=0}^{P} \beta_i \dot{v}_{n-i} \tag{3.3.1}$$

where n represents the index of current time-point, $n - i$ the index of time-point i steps before, v_n are the current estimates of time derivatives, $h_n = t_n - t_{n-1}$ coefficients dependent on h_n, h_{n-1}, h_{n-p+1}.

The formula (3.3.1) is called linear differentiation formula. It is said to be of order r if

$$\varepsilon_T = h_n\|\dot{v}(t_n) - \dot{v}_n\| = o(h^{r+1}) \tag{3.3.2}$$

where ε_T is the local truncation error introduced due to discretisation (Chua and Lin (1)). The simplest case of the formula (3.3.1) is the first-order Backward Euler formula

$$h_n\dot{v}_n = v_n - v_{n-1} \tag{3.3.3}$$

However, more accurate formulas such as second-order Shichman formula with variable step (Shichman (2))

$$h_n\dot{v}_n = \frac{2h_n + h_{n-1}}{h_n + h_{n-1}}v_n - \frac{h_n + h_{n-1}}{h_{n-1}}v_{n-1} +$$

$$\frac{h_n^2}{h_{n-1}(h_n + h_{n-1})}v_{n-2} \tag{3.3.4}$$

or variable-order variable-step backward differentiation formulas (BDF)

$$h_n\dot{v}_n = \sum_{i=0}^{P} \alpha_i v_{n-i} \tag{3.3.5}$$

are more desirable for circuit simulation due to their greater numerical efficiency and excellent stability properties.

It is useful to make distinction between the unknown current values of primary circuit variables v_n and the past values v_{n-i}, $i = 1, \ldots, P$ and rewrite the formula (3.3.1) as

$$h_n\dot{v}_n = \alpha_0 v_n + X_{v,n} \tag{3.3.6}$$

where $X_{v,n} \triangleq \sum_{i=1}^{P} (\alpha_i v_{n-i} + h_n\beta_i v_{n-i})$ contains only past information.

Application of (3.3.6) to the original equation set (3.2.1) yields a system of algebraic nonlinear equations for every discrete time point t_n

$$\tilde{f}(v_n) = 0 \tag{3.3.7}$$

where $\tilde{f}: R^N \to R^N$ is a continuous function of v_n.

Virtually all solution methods of (3.3.7) are based on the Newton-Raphson (NR) linearisation which is derived from Taylor expansion of (3.3.7) around the m-th estimate $v_n^{(m)}$ of the solution vector v_n.

$$\tilde{f}(v_n^{(m)} + \Delta v_n^{(m)}) = \tilde{f}(v_n^{(m)}) + J^{(m)}\Delta v_n^{(m)} + \ldots \tag{3.3.8}$$

where

$$\mathbf{J}^{(m)} = \frac{\partial f}{\partial v_n}\bigg|_{v_n = v_n^{(m)}} \varepsilon \ \ R^{N \times N} \text{ is the Jacobian matrix evaluated for } v_n^{(m)}.$$

If the terms of order higher than the first in (3.3.8) are rejected then the equation (3.3.7) reduces to its linearised estimate

$$\mathbf{J}^{(m)}\Delta v_n^{(m+1)} = b^{(m)} \tag{3.3.9}$$

where $b^{(m)} \triangleq - \tilde{f}(v_n^{(m)})$ and the solution $v_n^{(m+1)} \triangleq v_n^{(m)} + \Delta v_n^{(m)}$.

If $\|v_n^{(m+1)} - v_n\| < \|v_n^{(m)} - v_n\|$, $m = 0, 1, 2, \ldots$ then the repeated application (3.3.9) is a contraction mapping and the subsequent values $v_n^{(m)}$, $m = 0, 1, 2, \ldots$ converge to the solution v_n of (3.3.7). A simplified algorithm for the solution of (3.2.1) by a classical circuit simulator can now be summarised as follows.

Algorithm 3.1
{Comment: n is the time-point counter, m is NR iteration counter, ε is an arbitrarily small number, T is the maximum analysis time}

  n: $= 0$;   t: $= 0$;

  repeat

    n: $= n + 1$; t_n: $= t_{n-1} + h_n$;

    evaluate current value of differentiation operator α_0, $X_{v,n}$;
    predict values of unknown variables $v_n^{(0)}$
    as an initial guess for NR iterations;

    m: $= 0$;

    repeat

      m: $= m + 1$;

      set up the Jacobian $\mathbf{J}^{(m)}$ for current estimate of unknown variables

      $v_n^{(m)}$;

      solve linearised equation $\mathbf{J}^{(m)}\Delta v_n^{(m+1)} = b^{(m)}$

    until NR iterations converge i.e. $\|\Delta v_n^{(m+1)}\| < \varepsilon$

    select new step-size h_{n+1} on the basis of local truncation error estimation

  until $t_n > T$;

Companion models
Substitution of the time-derivatives by their algebraic approximations can be

interpreted in terms of circuit elements as substitution of capacitors and inductors by some resistive models for each time point $t = t_n$. Indeed, the application of differentiation formula (3.3.6) to the current-voltage relationship of a capacitor

$$i(t) = \frac{d}{dt} q(v(t)) = \frac{dq(v)}{dv} \cdot \frac{dv(t)}{dt} \tag{3.3.10}$$

yields for $t = tn$

$$i_n = \frac{\alpha_0}{h_n} q(v_n) + \frac{1}{h_n} X_{q,n} \tag{3.3.11}$$

For the purpose of Newton-Raphson iterations the equation (3.3.11) has to be converted to its linearised approximation by applying Taylor expansion around the m-th estimate of the unknown variable $v_n^{(m)}$, $m = 0, 1, \dots$

$$i_n^{(m+1)} = \frac{\alpha_0}{h_n} C(v_n^{(m)}) \Delta v_n^{(m+1)} + i_n^{(m)} \tag{3.3.12}$$

where

$$C(v_n^{(m)}) = \frac{dq(v_n)}{dv_n} \Bigg|_{v_n = v_n^{(m)}} \qquad \text{is voltage-dependent}$$

capacitance,

$$i_n^{(m)} = \frac{\alpha_0}{h_n} q(v_n^{(m)}) + \frac{1}{h_n} X_{q,n} \quad \text{is the m-th estimate of the capacitor}$$

current.

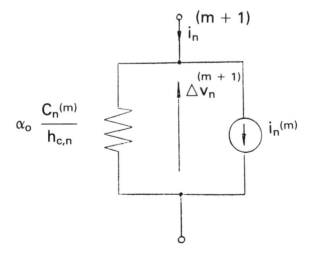

Fig. 3.1   *The companion model of capacitance*

The formula (3.3.12) is called the companion model of capacitance and it can be represented by a parallel connection of the conductance $\frac{\alpha_0}{h_n} C(v_n^{(m)})$ and the current source $i_n^{(m)}$ as illustrated in Fig. 3.1.

3.4 Standard solution of linear equations

Only a small fraction of the entries of $\mathbf{J}$ are nonzero. For $N > 500$, $\mathbf{J}$ has typically less than 2% of nonzero entries (Newton et al (4)).

In most circuit simulators the linearised equation (3.9) is solved by means of direct methods such as sparse Gaussian elimination or LU decomposition (Tewarson (5)). These methods have proven to be reliable and accurate. However, for large circuits the solution process can take a considerable amount of time even for powerful computers. Fig. 3.2 shows a typical CPU time required for a transient analysis of circuits of increasing size. The circuit size is measured in terms of number of equations N. For small circuits ($N < 30$) most of the solution time is spent on Jacobian build-up whereas the actual solution is very fast. However, when the circuit size grows an increasing part of the CP time is the solution time. As it can be seen in Fig. 3.2 for big circuits ($N > 500$) the solution time overwhelms the formulation time. The solution time has been measured to grow as $0(N\beta)$ where $1\cdot3 < \beta < 1\cdot6$ whereas the formulation time grows linearly with the number of circuit elements and, therefore for typical circuits, with the number of circuit equations.

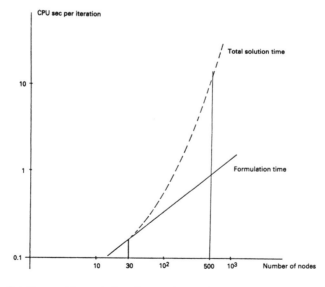

Fig. 3.2   *Solution and formulation time vs. increasing circuit size*

Also, the percentage of latent nodes, which have the voltages unchanging at a given time-point, increases with the circuit size. For typical circuits containing over 1000 MOSFETs, fewer than 15% of node voltages change significantly over a single time-step. Latency can be exploited by avoiding reevaluation of a vast proportion of Jacobian entries since for unchanging branch voltages the corresponding Jacobian entries from the previous Newton-Raphson iteration can be reused. This technique can be more efficiently implemented when the circuit equations are partitioned hierarchically and the solution process is also carried out in a hierarchical fashion (Zwolinski and Nichols (7)).

3.5 Relaxation methods

Relaxation (or iterative) methods can also be used for the solution of (3.2.1) in numerous ways. They can be implemented at all the three stages in the solution of (3.2.1) as shown in Fig. 3.3. Waveform relaxation (Lelarasmee et al (8)) is applied directly to the nonlinear differential equations (3.2.1). Nonlinear relaxation (Ortega and Rheinboldt (9)) is for solution of the algebraic nonlinear equations (3.3.7) resulting from application of differentiation formulas to (3.2.1). Linear relaxation is an iterative technique for solution of the linearised equation set (3.3.9).

Linear relaxation
Let the Jacobian matrix $\mathbf{J}$ be expressed as the matrix sum

$$\mathbf{J} = \mathbf{D} - \mathbf{E} - \mathbf{F} \tag{3.5.1}$$

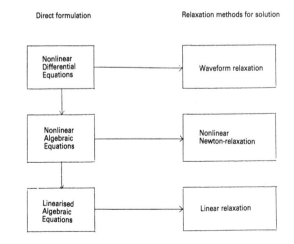

Fig. 3.3   *Implementation of relaxation-based techniques to circuit simulation*

where $\mathbf{D} = \text{diag} \{J_{11}, J_{22}, \ldots, J_{NN}\}$ and $\mathbf{E}_\varepsilon R^{N \times N}$ and $\mathbf{F}_\varepsilon R^{N \times N}$ are respectively strictly lower and upper triangular matrices. We assume $\mathbf{J}$ is nonsingular and all its diagonal entries are nonzero.

Then, the two basic linear relaxation methods for solution of (3.3.9) have the following form

Gauss-Jacobi (GJ)

$$\mathbf{D}v^{(k+1)} = (\mathbf{E} + \mathbf{F}) v^{(k)} + b \qquad (3.5.2a)$$

or $v^{(k+1)} = \mathbf{D}^{-1}(\mathbf{E} + \mathbf{F}) v^{(k)} + \mathbf{D}^{-1}b \triangleq \mathbf{P}_{GJ}v^{(k)} + \mathbf{D}^{-1}b \qquad (3.5.2b)$

Gauss-Seidel (GS)

$$(\mathbf{D} - \mathbf{E}) v^{(k+1)} = \mathbf{F}v^{(k)} + b \qquad (3.5.3a)$$

or $v^{(k+1)} = (\mathbf{D} - \mathbf{E})^{-1}\mathbf{F}v^{(k)} + (\mathbf{D} - \mathbf{E})^{-1}b \triangleq \mathbf{P}_{GS}v^{(k)} + (\mathbf{D} - \mathbf{E})^{-1}b$

$$(3.5.3b)$$

where $v^{(k)}$ is the solution estimate at the k-th iteration, the matrices $\mathbf{P}_{GJ} = \mathbf{D}^{-1}(\mathbf{E} + \mathbf{F})$ and $\mathbf{P}_{GS} = (\mathbf{D} - \mathbf{E})^{-1} \mathbf{F}$ are called the point Gauss-Jacobi matrix and the point Gauss-Seidel matrix correspondingly.

Theorem 3.1
The Gauss-Jacobi (or Gauss-Seidel) iterations converge for any initial guess $v^{(0)}$ if the eigenvalues λ_i, $i = 1, \ldots, N$ of $\mathbf{P}_{GJ}$ (or $\mathbf{P}_{GS}$) hold the condition

$$\max |\lambda_i| < 1 \qquad (3.5.4)$$
$$1 < i < N$$

It is difficult to check this condition in practice and more strict, but more computationally convenient, generally only sufficient conditions can be used instead.

Theorem 3.2
If $\mathbf{J}$ is strictly diagonally dominant i.e. if

$$|J_{ii}| > \sum_{\substack{j=1 \\ j \neq i}}^{N} |J_{ij}| \text{ for all } i = 1, \ldots, N$$

then the associated $\mathbf{P}_{GJ}$ and $\mathbf{P}_{GS}$ matrices are both convergent, i.e. they hold the condition (3.5.4).

In general, active circuits with controlled sources are unlikely to have diagonally dominant Jacobians since controlled sources introduce large off-diagonal entries. However, for circuits with grounded capacitors at every node a remarkable theorem can be proved.

Theorem 3.3

If a circuit can be described by nodal equation (3.2.2) and there is a grounded capacitor at every node then application of the linear differentiation formula (3.3.6) yields strictly diagonally dominant Jacobian provided the step-size h_n is sufficiently small.

Proof follows from the resulting linearised algebraic equation which has the form (3.2.2), (3.3.8), (3.3.11), (3.3.12)

$$\frac{\alpha_0}{h_n} C(v_n^{(m)}, u_n) + \frac{\partial q}{\partial v_n} (v_n^{(m)}, u_n)) \, \Delta V_n^{(m+1)} + i_n^{(m)}$$

$$+ \, q(v_n^{(m)}, u_n) = 0, \quad m = 0, 1, \ldots \tag{3.5.5}$$

For small stepsizes h_n the capacitive entries of Jacobian overwhelm the conductive entries introduced by the partial derivatives of q. The matrix C is strictly diagonally dominant and, therefore, there exists $h_{n,min}$ such that Jacobian is strictly diagonally dominant for $h_n < h_{n,min}$.

Nonlinear relaxation

Relaxation methods can also be used directly to solve nonlinear algebraic equation (3.3.7) and hence replace the solution of linearised equation (3.3.9). The appropriate Gauss-Jacobi and Gauss-Seidel algorithms can be devised as follows

Nonlinear Gauss-Jacobi algorithm:

{Comment: $v^{(0)}$ is the initial guess evaluated by prediction

            k   is the iteration index}

k: = 0;

repeat

    forall (i = 1, ..., N)

       solve $f_i(v_1^{(k-1)}, \ldots, v_i^{(k)}, \ldots, v_N^{(k-1)}) = 0$ for $v_i^{(k)}$ (3.5.6)

until $\|v^{(k)} - v^{(k-1)}\| < \varepsilon$;

The forall (i = 1, ..., N) statement specifies that the computations for all values of i can be carried out concurrently, i.e. they can be conveniently programmed for parallel computers.

    Nonlinear Gauss-Seidel algorithm:

k: = 0;

repeat

    foreach (i = 1, ..., N)

       solve $f_i(v_1^{(k)}, \ldots, v_i^{(k)}, \ldots, v_N^{(k-1)}) = 0$ for $v_i^{(k)}$ (3.5.7)

until $\|v^{(k)} - v^{(k-1)}\| < \varepsilon$;

The foreach ($i = 1, \ldots, N$) statement specifies the sequential order of computations as in the Gauss-Seidel algorithm the values of $v_1^{(k)}, \ldots, v_{i-1}^{(k)}$ must be known before the solution for $v_i^{(k)}$ can proceed.

Variants of Gauss-Seidel nonlinear relaxation are used extensively in the existing circuit simulators which rely on iterative methods. For example, the experimental circuit simulator HTRAN (Nichols and Kazmierski (11)) can outperform classical circuit simulators by combining nonlinear relaxation with multilevel hierarchical circuit description and exploitation of latency in inactive subcircuit clusters. A test simulation (Fig. 3.4) of an ECL 8-input multiplexer took less than 10 min CPU time on a VAX 11/750. The circuit was constructed from 4-input and 2-input multiplexers which in turn contained 2-input NOR gates (and 196 transistors altogether).

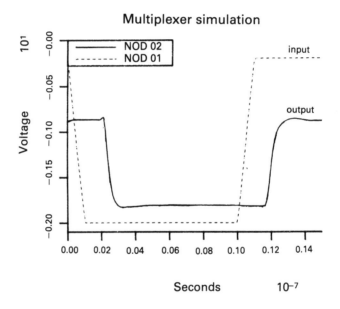

Fig. 3.4   *ECL multiplexer simulation*

Waveform relaxation
The idea of relaxation can also be applied to the original nodal equations in their differential form (3.2.2).

For point-GJ and point-GS waveform iterations the equation (3.2.2) is decomposed into N independent differential equations in one unknown. The relaxation process starts with an initial guess $v^{(0)}(t)$ in the given time interval $t_\varepsilon[0,T]$.

For the GJ waveform relaxation the equation (3.2.2) is transformed as following

$$C_{ii}(v_1^{(k-1)}, \ldots, v_i^{(k)}, \ldots, v_N^{(k-1)}, u) \dot{v}_i^{(k)} +$$

$$\sum_{\substack{j=1 \\ j \neq i}}^{N} C_{ij}(v_i^{(k-1)}, \ldots, v_i^{(k)}, \ldots, v_N^{(k-1)}, u)\dot{v}_j^{(k-1)}$$

$$+ q_i(v_1^{(k-1)}, \ldots, v_i^{(k)}, \ldots, v_N^{(k-1)}, u) = 0,$$

$$i = 1, \ldots, N \qquad (3.5.8)$$

where $k = 1, 2, \ldots$ is the iteration index and $v_i^{(k)}$ is the unknown waveform. For the GS relaxation, the waveform solution obtained by solving one decomposed system is immediately used to update the approximate waveforms of the other equations. The transformation of equation (3.2.2) has the following form

$$\sum_{j=1}^{i} C_{ij}(v_1^{(k)}, \ldots, v_i^{(k)}, \ldots, v_N^{(k-1)}, u) \dot{v}_j^{(k)}$$

$$+ \sum_{j=i+1}^{N} C_{ij}(v_1^{(k)}, \ldots, v_i^{(k)}, \ldots, v_N^{(k-1)}, u) \dot{v}_j^{(k-1)}$$

$$+ q_i(v_1^{(k)}, \ldots, v_i^{(k)}, \ldots, v_N^{(k-1)}, u) = 0, \quad i = 1, \ldots, N \quad (3.5.9)$$

The iterations are carried out according to the following two basic algorithms.

Gauss-Jacobi waveform relaxation algorithm:

$k := 0;$ forall $(t_\varepsilon[0,T]) \, v^{(0)}(t) := v(0);$

repeat

   $k := k + 1;$

   forall $(i = 1, \ldots, N)$

      solve equation (3.5.8) for $v_i^{(k)}(t)$

until $\max_{t_\varepsilon 0, T} \|v^{(k)}(t) - v^{(k-1)}(t)\| < \varepsilon;$

Gauss-Seidel waveform relaxation algorithm:

$k := 0;$ forall $(t_\varepsilon 0, T) \, v^{(0)}(t) := v(0);$

repeat

   $k := k + 1$

   foreach $(i = 1, \ldots, N)$

      solve equation (3.5.9) for $v_i^{(k)}(t)$

until $\max_{t_\varepsilon 0, T} \|v^{(k)}(t) - v^{(k-1)}(t)\| < \varepsilon;$

Convergence of nonlinear and waveform relaxation methods
In general, convergence is achieved if an iteration process is contraction mapping, i.e. if

$$\left\| v_n^{(k)} - v_n \right\| < \left\| v_n^{(k-1)} - v_n \right\|$$ (3.5.10)

for nonlinear relaxation and

$$\max_{t \in [0,T]} \left\| v^{(k)}(t) - v(t) \right\| < \max_{t \in [0,T]} \left\| v^{(k-1)}(t) - v(t) \right\|$$ (3.5.11)

for waveform relaxation. The conditions under which nonlinear and waveform relaxations converge are analogous to those for linear relaxation. In particular, if there is a grounded capacitor at each node, theorems similar to the Theorem 3.3 apply, in other words, one can obtain a convergent iterative process provided the integration step-size is appropriately small.

3.6 Conclusion

The computational advantages of relaxation become more evident for very large circuits as the total number of operations grows proportionally with the circuit size. This makes circuit simulation for VLSI a realistic possibility which can be exploited seriously in the process of integrated circuit design. However, can the relaxation based simulators really outperform classical circuit analysis programs in terms of accuracy and reliability in dealing with a wide range of complex electronic circuits? The competition is open and the question is yet to be answered.

References

1   CHUA, L., LIN, P.: 'Computer-Aided Design of Electronic Circuits', Prentice-Hall Inc., Englewood Cliffs, New Jersey, 1975
2   SHICHMAN, H.: 'Integration System of a Nonlinear Network Analysis Program', IEEE Trans. on Circuit Theory, v. CT-17, 378–386, Aug 1970
3   VAN BOKHOVEN, W. M. G.: 'Linear Implicit Differentiation Formulas of Variable Step and Order', IEEE Trans. on Circuits and Systems, v. CAS-22, 109–115, February 1975
4   NEWTON, A. R., SANGIOVANNI-VENCENTELLI, A. L.: Relaxation-Based Electrical Simulation, IEEE Trans. on Computer Aided Design, v. CAD-3, 308–331, October 1984
5   TEWARSON, R. P.: 'Sparse Matrices', Academic Press, New York, London, 1973
6   RABBAT, N. B. G., SANGIOVANNI-VINCENTELLI, A. L., HSIEH, H. Y.: 'A Multilevel Newton Algorithm with Macromodelling and Latency for the Analysis of Large-Scale Nonliner Circuits in the Time-Domain', IEEE Trans. on Circuits and Systems, v. CAS-26, 733–741, September 1979
7   ZWOLINSKI, M., NICHOLS, G. G.: 'The Design of an Hierarchical Circuit-Level Simulator', Proceedings of Electronic Design Automation Conf., Warwick, UK, March 1984

8 LELARASMEE, E., SANGIOVANNI-VINCENTELLI, A. L., RUHELLI, A.: 'The Waveform Relaxation Method for the Time-Domain Analysis of Large-Scale Integrated Circuits, IEEE Trans. on Computer-Aided Design, v. CAD-1, 131–145, August 1982
9 ORTEGA, J. M., RHEINBOLDT, W. C.: 'Iterative Solution of Nonlinear Equations in Several Variables', Academic Press, New York, 1970
10 VARGA, J.: 'Matrix Iterative Analysis', Prentice Hall, Englewood Cliffs, New Jersey, 1962
11 NICHOLS, K. G., KAZMIERSKI, T. J.: 'Simulation Inside Out', Silicon Design, v. 2, no. 11, 22–44, November 1985

Chapter 4

Logic simulation algorithms and techniques

4.1 Introduction

In the design of digital circuits, simulation is used extensively in all phases of the design process; for example, in the initial phase of conceptual design, simulation is used to verify that a given configuration of switching elements, i.e. gates, flip-flops etc. realise some logic function. As a design progresses simulation will be used to examine the dynamic characteristics of the circuit with the objective of detecting any switching anomalies which exist in the circuit. Simulation is also used to determine the fault coverage of sets of test vectors which will be used to determine which devices are fault free after fabrication. Design alternatives, which may enhance the testability of the circuit or improve its performance can also be examined using simulation.

The original gate level simulators were, essentially, logic level simulators which operated upon sets of Boolean Equations and, hence, assumed that the simulation primitives, namely AND, OR, NOR and NAND gates, were ideal switching elements (zero delay). Present day simulators, however, use a wider range of primitives which include higher level functions, for example, flip-flops, registers, counters, PLA's, RAM's and ROM's etc. and are referred to as Functional or Block Level Simulators. The simulation primitives are also modelled more realistically by including delay parameters in the models, for example, rise and fall times, propagation delays and inertial delays; furthermore, the signal value assigned to the output of a primitive is no longer restricted to either a logic 1 or logic 0, but may be defined as rising, falling, in a 'high-impedance' state or in an undefined or unknown state. More recently, however, the behaviour of subfunctions within a circuit have been described by using special purpose high level languages which are keyword driven; arithmetic, logical and storage functions being implemented with the arithmetic and logical operators and assignment statements used in high level programming languages. Simulators which operate on circuits described in this way are usually referred to as Behavioural Level Simulators.

4.2 Simulation process (Breuer and Friedman (1))

The process of simulating a digital circuit comprises creating a model of the circuit, exercising the model by applying a set of input stimuli, which represent the driving waveforms for the circuit, and finally predicting the response of the model to the input stimuli; the response of the circuit is usually displayed as a plot of signal values at selected nodes in the circuit as a function of time.

4.2.1 Modelling a digital circuit for simulation

In order to model a digital circuit for simulation it is necessary, first, to describe the circuit to the computer in terms of the available simulation primitives and their interconnections. In general each simulator has associated with it a special purpose circuit description language comprising statements which define the types of primitives and their interconnectivity, described implicitly by the list of input and output signals associated with the primitive (the assumption which is made is that if a given signal name is in the input list to a primitive, a connection exists between that input and the source of the signal). The circuit description language may also contain statements which describe the various delay parameters associated with the primitives. Furthermore there will be facilities in the language to describe the driving waveforms to the circuit.

The syntax of the input description language is subsequently checked by the 'preprocessor module' of the simulator which scans the circuit description for possible errors; for example, illegal primitive types, non-unique signal names etc. The preprocessor may also have a macro expansion capability which facilitates the detailed description of circuits containing many repetitive blocks.

A model of the circuit is generated subsequently from the syntactically correct circuit description by the 'model compiler' module in the simulator. If the circuit description is compiled into a sequence of computer instructions, which will predict the circuit response to a given set of inputs, when executed by the computer, the simulator is referred to as a Compiled Code Simulator. However, if the circuit description is translated into a set of interlinked data tables, which are operated upon by a separate simulation program module called the 'simulator executive', the simulator is said to be Table Driven. An example of the Compiled Code and Table Driven Models of a circuit, is shown in Fig. 4.1; in the case of the Table Driven Model only the main tables and associated pointers are shown. The main advantage offered by the Compiled Code Simulator is its speed of operation; however, its main disadvantages are first, circuit modifications cannot be made easily, requiring the recompilation of the simulation model; second, in order to avoid erroneous simulation results it is necessary to 'level' the circuit description before the Compiled Code Model is created; this ensures that all output signal values are updated before they are used as inputs to successor gates. Consequently, this type of simulator

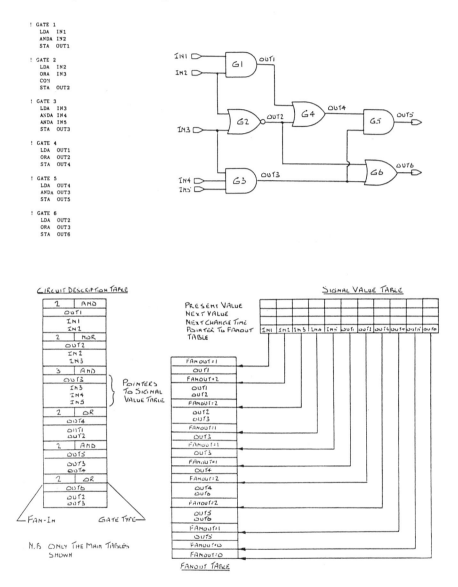

Fig. 4.1   *Compiled code and table driven circuit model*

is more suitable for evaluating combinational circuits in which the switching primitives are considered to be ideal. However, the Table Driven Simulator, although slower than the Compiled Code Simulator, has many advantages. First, the algorithms used in the Table Driven Simulator automatically ensure the correct sequencing of gate evaluations, irrespective of the order in which the gates appear in the circuit description; second, the simulation primitives may include delay parameters and also signal transitions are not limited to two-values. Furthermore, since each signal source has its own fanout list the technique of 'Selective Trace' simulation can be used to improve the simulation efficiency of large circuits. In the technique of Selective Trace Simulation, when a given signal changes value the immediate effect upon the circuit can be determined by examining only those gates in the fanout list of the signal which changed value. This forward referencing of signal changes cannot be incorporated in the Compiled Code Simulator, hence in order to evaluate a given change in a signal value all gates in the circuit must be examined.

4.2.2 Basic Simulation algorithm

The process of simulating the function of a digital circuit comprises

(*a*) Evaluating the effect of signal changes in the circuit.
(*b*) Scheduling events (signal changes) in the correct time sequence.

In this way the output response of the circuit to a given set of input signal changes can be predicted; this is the main function performed by the 'simulator executive' module in the simulator.

The basic algorithm used to simulate a logic circuit is outlined below, assuming that the primitives have delay parameters associated with them and signal transitions are represented as zero, one and unknown.

(*a*) At time zero, set all gate outputs to an unknown value.
(*b*) Apply the input stimuli (driving waveforms) to the circuit.
(*c*) Using the technique of Selective Trace Simulation evaluate the effect of the changes in logic value, which have occurred in the circuit. (At time zero, simply determine the effect of applying the input stimuli.)
(*d*) Schedule, in an event queue, the times when signal changes will occur at gate outputs or in the driving waveforms.
(*e*) If the duration of the simulation has expired – terminate the simulation run; else, set the simulation clock to the next event in sequence, in the time queue.
(*f*) Implement the changes in signal values occurring at this time – go to (*c*).

Evaluation of Logic Changes in a Circuit: In general, several logic changes (events) will occur simultaneously in a circuit; however, the computer cannot process events in parallel. To contend with this situation time is momentarily 'frozen' within the simulator and the changes in logic value are updated one at a time; thereafter, for each signal source, in turn, which has changed value, the subsequent logic value on the outputs of the gates in their fanout lists are determined, to evaluate the effect of the change in logic value upon the circuit. Since all signal updates are performed before individual gate evaluation procedures are implemented, then any gate evaluation performed will include the effect of all the changes to the input signals on a gate which have occurred at that time.

The procedure (Newton (2)) for evaluating the effect of the changes in logic value, which have occurred at some instant in time, in a circuit is summarised below.

FOR (Each gate G which has changed value and time T) *DO*
FOR (Each gate G_n in the fanout list of G) *DO*
BEGIN
Obtain the input signal values to G_n;
Compute the new output value of G_n;
IF (Present output value $\neq$ New output value)
    THEN (Schedule the gate output change in the Event Queue at sometime
        $T + \triangle t$)
END

Event scheduling during simulation: Since simulation is the calculation of logic values as a function of time, an important aspect of the simulation process is the scheduling of events (gate changes) in the correct temporal sequence.

The mechanism for scheduling events can be implemented using the technique of 'Next Event List Processing' which permits events to occur asynchronously in the circuit, or by a 'Time Mapping Algorithm' (Ulrich (3)) which restricts the occurrence of events to some fixed increment of time.

In the 'Next Event List Processing' Technique a simple list structure is created in which events are ordered with increasing time. Whenever an event is to be scheduled the list must be scanned so that the event is placed in the correct sequence in the time queue, since all delays in a circuit are not identical, an event later in time, may cause a future event to occur which is earlier in time than an event already placed in the time queue. If the time queue or list is long, inferring that there is a lot of activity in the circuit, a large amount of CPU time is wasted by processing the list, this is the major disadvantage of this technique. However, its advantages are that it is simple to implement, there are no restrictions upon times when events can occur and the advancement of time is accomplished by jumping from one scheduled event to another thus avoiding periods of inactivity in the circuit.

The Time Mapping Technique is an algorithmic method for scheduling events during simulation. In this technique a circular list ($\triangle t$-loop) is set up as shown in Fig. 4.2(a), each section in the loop represents some increment in time, $\triangle t$, which is determined at the beginning of each simulation run from the delays in the circuit; $\triangle t$ is the greatest common denominator of the shortest and longest delays in the circuit, all delays are subsequently expressed in terms of $\triangle t$, L is subsequently assigned the value of the maximum element delay in the circuit. Each sector in the $\triangle t$-loop contains either a signal name or a pointer to a list of signals which are going to change state at a particular instant in time; for example, as shown in Fig. 4.2(a) it is assumed that signals A, B and C are going to change state at $2 \times \triangle t$, $3 \times \triangle t$ and $6 \times \triangle t$ respectively. During a simulation run the program scans each sector in the $\triangle t$-loop to determine if an event is going to take place in which case time is temporarily 'frozen' and the event is simulated. The outcome of the event is then scheduled into the appropriate time slot in the $\triangle$t-loop simply by adding the gate delay, which is an integer multiple of $\triangle$t's, on to the current position of the pointer in the $\triangle$t-loop. The $\triangle$t-loop method of event scheduling thus removes the need to scan a list of future events in order to schedule an event in the proper time sequence, which was the major disadvantage of the list processing technique when used in a circuit in which there was a lot of activity. However, in a circuit in which there is little activity, the $\triangle$t-loop method of event scheduling becomes inefficient since all sectors of the $\triangle$t-loop must be scanned regardless of whether or not they contain an event. For example, with reference to Fig. 4.2(a), if it is assumed that event B has just been simulated then, in order to get to the next event C, the intermediate empty sector must be scanned, since time can only be advanced in increments of $\triangle$t. This situation is aggravated when a relatively inactive circuit contains several very long delays in comparison to the other delays in the circuit, since a large timing wheel will be generated which contains many small slots. The problem of efficient event scheduling in circuits with long delays can be resolved if two timing loops are used as shown in Fig. 4.2(b), the increment in the first loop is still $\triangle$t, whilst that in the second loop is L, the length of the $\triangle$t-loop. After each revolution of the $\triangle$t-loop, the pointer to the L-loop is incremented by one and the events found in the appropriate sector (ie occurring at nL + m$\triangle$t, where m$\triangle$t<L) are placed in the $\triangle$t-Loop; events in a given sector of the L-Loop need not be ordered since this will be done automatically when they are placed in the $\triangle$t-loop.

Accuracy of simulation results: (Breuer and Friedman (1)) Since simulation is the calculation of signal value changes as a function of time, the accuracy of these calculations depends upon how closely the delays included in the gate models simulate the delays which exist in the physical circuit. The delays which are usually modelled in a simulator are,

(*a*) *Transport delay:* Each gate or connection that a signal passes through in a circuit must introduce some time delay, which is represented as a pure delay block attached to the output of an ideal logic gate.

(*b*) *Inertial delay:* If the width of a dominant input pulse to a gate is very short it will not force the gate to switch. The minimum time that a logic change must remain on an input before the gate responds is called the Inertial Delay Time of the gate. If the pulse width is greater than the Inertial Delay Time, the effect of the delay is similar to that of the Transport Delay, if not the input pulse is suppressed.

(*c*) *Rise and fall delays:* When a gate output changes value it does not affect the change immediately due to capacitances in the circuit, which must be charged or discharged. In general, rise and fall times have different values (of the order of four to one) resulting in an elongation of the output pulse.

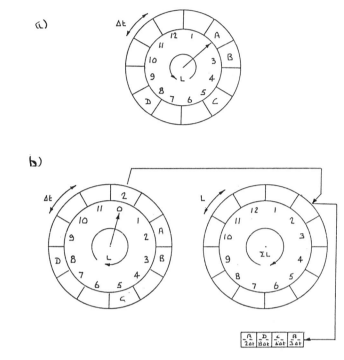

Fig. 4.2   *Next event scheduling algorithm*

(d) *Ambiguity delay:* This delay model was introduced to account for the spread in delay parameters of the actual devices used in implementing a given circuit. In modelling Ambiguity Delays, minimum and maximum values are assigned to a delay parameter which defines a time window when it is uncertain that, for example, a gate transition has occurred or not; all that can be stated is that the transition would not have occurred before the minimum time has elapsed but will have occurred before the maximum time has elapsed. The use of ambiguity delays can produce very pessimistic results.

(e) *Load dependent delay:* This delay associated with a gate is a function of the number of gates where a given gate output is used as an input. This delay parameter can be derived automatically by the simulator from the fanout listings of the gates in the circuit.

The accuracy of the simulation results can also be improved by increasing the number of values used to define the output state of a gate. The early simulators modelled a gate as a two-state switching element and had many limitations, for example, it inferred that at time zero a decision had to be made upon the most likely logic value which would exist on the output of the gates in the circuit. This limitation was overcome by introducing a third value, namely an X-value which defined the output of a gate as being 'unknown'. The third value was also used to describe the output of a gate which was in transition and could be used to trap hazard conditions on gates. Furthermore, when modelling ambiguity delays it is necessary to differentiate a '0 to 1' and '1 to 0' transition on a gate output thus requiring a five-valued simulation model. It has been reported that some simulators use between a nine- and a fifteen-values during simulation, however, the improved accuracy of the simulation results, is to some extent offset by the increase in CPU time required to evaluate the complex gate models.

4.2.3 *Gate level simulation anomalies* (Holt and Hutchings (4), Sherwood (5))

A characteristic of the switching primitives which make gate level simulation efficient for relatively large circuits, is that the primitives are considered to be unilateral switching elements, i.e. they have well defined inputs and outputs with signal propagation occurring in one direction only. However, with the advent of LSI and the introduction of the 'transmission gate' as a switching primitive it is necessary to model an element which not only has an output logic value which has a 'dynamic' state but also has the ability to exhibit a bilateral switching characteristic.

In order to model the dynamic state of a transmission gate, which occurs when the signal on the control gate terminal is a logic 0, a fourth logic value had to be introduced, namely, the high impedance value, Z, which would indicate that the output of the transmission gate was not 'driven'. However, this high impedance state, when used as an input to a conventional gate, was considered, usually, to be an unknown value and hence produced some pessimistic results. The description of the high impedance state was subsequently redefined in terms of a 'high-impedance-one' (Z1), a 'high-impedance-zero' (Z0) and a 'high-impedance-don't-know' (ZX), which on input to conventional gates were considered as one, zero and don't-know respectively. A time-out parameter was also included in the transmission gate model, after which time a Z1 or Z0 value would decay to a ZX state, indicating that when the control gate terminal is zero the known state will not remain there indefinitely, simulating the decay of charge of the node capacitance of a transmission gate.

Wired logic is also another function which must be modelled regularly in LSI circuits. With wired logic the outputs of several transmission gates are connected to a single node; in order to determine the logic value of the node when some transmission gates are switched on and others off, it is necessary to introduce a 'phantom' or 'consensus' node into the circuit. The 'phantom' or 'consensus' node is automatically inserted by the simulation program in order to simulate the wired −OR or multiplexor function.

The modelling of devices which exhibit a bilateral switching characteristic is also troublesome in a simulator which is oriented towards unilateral switching elements. One technique which has been used to model a bilateral transmission gate is shown in Fig. 4.3. The bilateral transmission gate is automatically replaced by two unilateral transmission gates and two phantom nodes, which are used to simulate the connection of the bilateral transmission gate to the other transmission gates in the circuit.

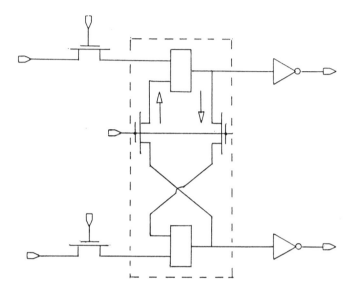

Fig. 4.3   *Model of a bilateral transmission gate*

4.3 Behavioural modelling of gate level functions (Robson (6), Noon (7))

The individual models within the simulator are written as procedures or subroutines which cannot be modified in any way by the user; in some situations this can be a disadvantage when the designer wants to simulate a function which either does not exist or is slightly different from the elements in the model library. This difficulty has been overcome by the introduction of Behavioural or Functional Modelling Languages.

These modelling languages are very similar to some high level programming languages, for example, PASCAL. The basic set of actions used to describe a function in a Behavioural Modelling Language are outlined below.

(*a*) *CAUSE:* This is an event which results in a function being evaluated, it is similar to an input to a gate changing value causing the gate function to be evaluated. The event can either be a simple signal change for example, a clock rising or falling or it may be the 'event' of a Boolean Expression becoming TRUE. This part of the statement is characterised by the keyword *WHEN*.

(*b*) *CONDITION:* This defines a controlling set of conditions which must exist when the event (CAUSE) occurred before some action can be performed. The Conditional part of the statement is prefixed by the keyword *IF.*

(*c*) *EFFECT:* The 'Effect' is the action or function to be performed when a given event occurs and the controlling conditions are valid. The Effect part of the statement is prefixed by the keywords *MAKE/DO.*

(*d*) *TIME:* This part of the statement is to account for delays in a function and is prefixed by the keywords *WITHIN/AFTER.*

(*e*) *TERMINATOR:* This is a nullifying condition which inhibits the Effect part of the statement from being performed, although the Cause and Conditional parts of statement would permit the function to be performed. The Terminator is used to simulate the action of say, Reset signals, and is introduced by the keyword *UNLESS.*

A basic behavioural statement(8) would take the form of

```
WHEN <        >    IF <        >
     MAKE/DO <      > WITHIN/AFTER<        >
     UNLESS <        >;
```

The above statement would also contain a Header Block defining the input and output signals to the block.

As a simple example of the use of a Behavioural Modelling Language consider the part description of a transmission gate shown below (Holt and Hutchings (4)).

```
EXAMPLE (Transmission Gate)
     WHEN GATE = 1 THEN
     IF SOURCE_STRENGTH > DRAIN_STRENGTH
        MAKE DRAIN = SOURCE WITHIN 20 nsecs ELSE
     IF (SOURCE_STRENGTH = DRAIN_STRENGTH)
        AND (SOURCE ≠ DRAIN)
        MAKE DRAIN = UNKNOWN WITHIN 20 nsecs

               .
               .
               .
        ETC.
```

The advantages of using a Behavioural Modelling Language to describe functions are that it provides a readable description of how a block operates or functions and checks can readily be incorporated to detect illegal values in the operation of a function. Behavioural Modelling Languages also allow analogue functions to be described, for example, D/A converters and Schmitt Triggers, and simulated in a gate level simulation environment with normal switching elements. The use of a modelling language also permits the descriptions of digital systems at higher levels of abstraction, this permits a more efficient simulation of complex systems since the evaluation of a single behavioural statement may be equivalent to the simulation of say one hundred gate changes at the lower level. The ability to describe the function of large logic blocks by means of a Behavioural Modelling Language is a prerequisite to top-down design methods, since in this technique the system is partitioned into several subfunctions whose operation is described using a Behavioural Modelling Language, thereafter each block in turn is decomposed into lower level functions; at each stage of the decomposition the new representations of a block can be simulated with the other subfunctions described at the higher level, this enables the integrity of the decomposition to be checked as well as the interfaces between the blocks.

The capability, of simulating simple functions described using a Behavioural Modelling Language, can readily be incorporated into a basic gate level simulator by including a 'command interpreter' function, in addition to the basic gate simulation routines, together with a Behavioural Description Table which contains an internal representation of the function described in the Behavioural Modelling Language. It should be noted, that although a Behavioural Description of a function may look like a small computer program it differs from it in one major aspect, that is, the statements are not executed sequentially but driven by events. The simulation of these functions is activated, as a normal gate, by changes in the logic value of their inputs.

4.4 Conclusions

The basic algorithms and data structures for a gate level simulator have been outlined. These can be readily modified to operate on more complex gate models which describe output transitions using more than 3-values or incorporate a more accurate delay model which would enable hazard trapping to be performed. The ability to extend the range of functions which can be simulated, through the use of a behavioural description language, greatly enhances the overall capability of the simulator, allowing it to simulate analogue devices in a digital environment and also functional blocks which have not yet been decomposed into gate level primitives.

At present, although simulation is widely used in industry as the main vehicle for verifying circuit designs, it has two major limitations. First, with present day circuit complexities, gate level simulators use a vast amount of CPU time, this overhead, however, has been reduced, to some extent, by the development of special purpose simulation engines which are at least twenty times faster than simulators run on a general purpose computer; these special purpose engines are discussed in Chapter 15. Second, the verification process performed using simulation is incomplete, since the simulation results will only indicate that the circuit operates correctly for the data inputs applied at that time. In order to obtain a complete verification of a circuit more formal methods of verification, similar to those used to prove the correctness of computer programs, are being examined. These techniques are discussed in Chapter 14. However, despite its limitations it is considered that gate level simulators will be used for many years to come, since formal verification methods are still in their infancy and the use of special purpose engines only offer a speed advantage over gate level simulators if a large amount of input data is to be simulated.

4.5 References

1   BREUER, M. A., FRIEDMAN, A. D.: 'Diagnosis and Reliable Design of Digital Systems', Pitman, 1977
2   NEWTON, A. R.: 'Timing, Logic and Mixed-Mode Simulation for Large MOS Integrated Circuits', Nato Advanced Study Institute Course Notes, Urbino, 1980
3   ULRICH, E. G.: 'Exclusive Simulation of Activity in Digital Networks', Communications of the ACM, February 1969, Vol. **12**, No. 2, pp. 102–110
4   HOLT, D., HUTCHINGS, D., 'A MOS/LSI Oriented Logic Simulator', 18th Design Automation Conference Proceedings, June 1981, pp. 280–287
5   SHERWOOD, W., 'A MOS Modelling Technique for 4-State True-Value Hierarchical Logic Simulation', 18th Design Automation Conference Proceedings, June 1981, pp. 775–785
6   ROBSON, G., 'Logic Design Using Behavioural Models', VLSI Design, January 1984, pp. 36–44
7   NOON, W. A., 'A Design Verification and Logic Validation System', 14th Design Automation Conference Proceedings, June 1977, pp. 362–368

Chapter 5

Testing VLSI circuits

K. Baker and G. Russell

5.1 Introduction

One major issue in the design of VLSI circuits is that of testing the devices after fabrication as wafers, ICs and systems in the field. The major problems with testing VLSI Circuits (1) are,

(1) Testing costs which are proportional to testing time are increasing rapidly with VLSI circuit complexity.

(2) Test pattern generation time is increasing with circuit complexity, this however, is offset slightly by the improved performance and reduction in the cost of computer systems.

(3) The amount of data, that is the input patterns and output responses, required to test a circuit is expanding more rapidly than either the testing or test pattern generation costs.

Attempts to reduce some of these costs have been made by developing more efficient test generation techniques which include improved algorithms for gate level test generation or algorithms which enable tests to be generated from high level descriptions of the circuit. Several design strategies (2) have also evolved which make complex circuits more testable by including hardware which, at test time permits the circuit to be reconfigured into less complex subfunctions or permits the circuit to adopt a self-test mode of operation.

It has been established that testing VLSI circuits is costly, but why is it necessary to test the circuits in the first instance, if the design has been simulated and shown to be correct? The faulty circuits result from a wide range of defects which occur in the fabrication (3) for example, pin-holes in gate oxide, shorted or open interconnect lines, contact hole defects, crystalline defects on the wafer etc., and also some design faults which have not shown up during simulation, since simulation is an incomplete process based on a abstracted model.

The process of test pattern generation is concerned with deriving input patterns (test vectors) which will produce a different output response from a circuit depending upon whether the circuit is faulty or fault free. However, the process of test generation not only includes the derivation of test vectors but also the closely related topics of fault modelling and fault simulation, thus before describing some test generation algorithms these two topics will be discussed briefly.

Testing is basically an economics issue. If a zero-defect IC could be produced then production testing would be redundant, hence the need is to test ICs at the lowest cost consistent with necessary quality levels. In order to achieve low test costs, but retain high product quality it is necessary to use design for testability. The nature of the CAT (computer aided test) tools used is dependent on the design for testability (DFT) methods used in that design group. DFT and CAT tools are intimately linked, the wrong tools in the strategy will result in poorly testable devices and high test costs. For this reason it is important to have a clear notion of DFT before any discussion of CAT tools. In this chapter, three approaches to DFT are considered; Ad-Hoc, classical structured DFT and neo-structured DFT (4,5,6).

5.1.1 What is the IC test problem?

In Fig. 5.1 is a graph with a family of curves, on the X-axis of the graph is the test coverage for a particular functional test, on the Y-axis is the quality of the supplied ICs (e.g. number of defective devices shipped as good devices), on

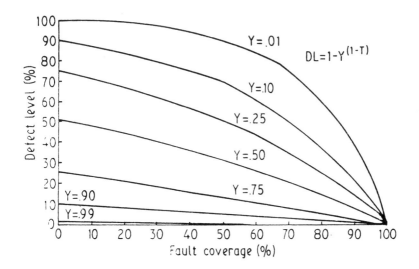

Fig. 5.1 *Defect level as a function of fault coverage*

the graph are plotted a set of curves that represent potential yields for that IC (7).

This graph shows that for a high-yield process the level of fault coverage required to achieve good quality levels is quite low. This in turn implies that neither great effort is needed to develop a high coverage test pattern nor is there need for intrusive design for testability schemes. However, at the other extreme where yield is low then the need for high fault coverage is paramount to achieve high quality. In practice, this means that either a costly manually generated test program or extensive design for testability will be needed. The latter option will of course have its own impact on the yield of the IC.

5.2 Design for testability

It has been mentioned in the introduction that DFT strategy and CAT tools are very closely related. For a DFT strategy to achieve its aim CAT tools must be carefully selected. In many cases suitable tools are not commercially available and internal developments are essential. In testing, if no other area in VLSI CAD, a strong design methodology supported by the necessary CAD tools can make or break the successful engineering of VLSI devices.

Historically, developments in DFT techniques have been pioneered by the large mainframe computer manufacturers, in particular, IBM have led developments in this area (8). In the following sections, the basic approaches to DFT are outlined.

5.2.1 Ad-Hoc methods
When testing, problems are first noted by a design group, probably after a long automatic test generation (ATG) or fault simulation run, the initial aim of the designer has been to find a 'band-aid' solution, an Ad-Hoc solution for a particular problem. To aid such design groups many papers and reviews of these techniques have been published over the years (8, 9). Over the course of the years it became obvious that these Ad-Hoc solutions were not the complete solution, they solved a problem but only after a great deal of thought or computer time was employed on a fruitless search. To reduce the time spent using valuable expertise, or expensive computers, a number of testability analysis tools were developed (10, 11). These tools all attempt to characterise the design in terms of controllability and observability. Controllability is the ease by which the internal nodes and states of the logic can be set or reset, observability is the reverse, the ease by which sensitive paths can be set up to monitor the internal nodes and states of the logic. Armed with these definitions analysis based on internal signal probabilities, node counts and other measures have been created.

For a brief period these tools enjoyed popularity within the industry, although many were unhappy with their performance. In depth studies of one of the most popular, SCOAP, have shown that such tools are very poor at assessing absolute testability and should be used with great care (12). Furthermore, it is noted that analysis based on structural descriptions are applied too late in the design cycle to be effective. However, even if they had been reasonably accurate, the applications of testability analysis would have been quite limited because it can never guarantee automatic test generation.

The main application of testability analysis would seem to be in the area of training and education concerning DFT.

5.2.2 Classical structured methods

It was observed by many in CAT research projects during the sixties and seventies that although ATPG for combinational logic was a reasonably effective tool, this was not true for sequential logic. In fact, ATG was only effective when heavily supported by the designers, human expertise and fault simulation tools. This suggested to some that complete ATPG would only be possible if the only combinational logic under test was the type of random logic encountered in TTL and gate array logic, this led to the proposal of structured DFT schemes based on scan path or LSSD (Level Sensitive Scan Design) (8).

In scan path or LSSD, the designer is forced to design the logic in a synchronous manner, each of the internal storage elements having been based on either D-type flip-flops or master-slave latch pairs. These can then be modified such that the flip-flops operate in two modes:

(1) Normal mode
   Normal D input is enabled at the latch/flip-flop, when clocked.

(2) Scan mode
   A secondary scan input is enabled at the latch/flip/flop when clocked.

Each of these secondary scan inputs is linked to the Q output of the last flip-flop to form a shift register chain. This can be used to control and observe all the internal sequential nodes of the circuit, see Fig. 5.2. To test the circuit it is only necessary to test that the scan path operates as a shift register and then test the combinational logic via the scan path. This is a simple operation that is split into three phases:

(1) Phase one
   In this phase the test equipment loads the test vector by shifting it along the scan path shift register using the scan input. The scan path is in shift mode for this operation.

(2) Phase two
   In this phase the combinational logic is tested by running one clock cycle in the normal mode, hence the scan path captures the test result vector.

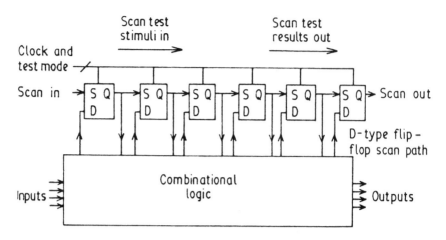

Fig. 5.2   *An idealised scan-path circuit*

(3) Phase three
   In this phase the test equipment unloads the result vector by shifting it along the scan path shift register to be sampled at the scan output. The scan path is in shift mode for this operation.

This method has been shown to work well, provided the designers can accept the constraints of scan design, and some simple ATPG tools are available. However, in designs where the gate count is low (1000 gates) or performance is vital then the overhead can be unacceptable.

Design methodologies for scan design are most advantageous where all of the advantages of synchronous design has been utilised, ATPG, simple, fast simulation tools, race-free logic. A good example of such a strategy is the UK5000 gate array, where the array has been designed with the scan path flip-flops as an integral part of the logic (13).

5.2.3 Neo-structured methods

The true introduction of VLSI design brings many changes when compared to the TTL or gate array design style. In true VLSI, regular structures are extensively used to improve designer productivity and system performance. For example, in most VLSI circuits at least one of the following regular structures can be found:

(1) RAMs, ROMs or PLAs
(2) Systolic Arrays
(3) Iterative Arrays
(4) Gate Matrices
(5) NMOS or Dynamic CMOS Switch Networks
(6) 'Strip' Logic

To ensure complete testability it is necessary to have a DFT strategy to cover each of these logic types. For the reasons given below it is obvious that simple scan path methods will not work. It would seem that it will be necessary to mix serial, and parallel access with built-in self-test (BIST) methods to achieve acceptable testability. However, if this is to be achieved, DFT will need to be carefully planned and very much earlier in the design when the floorplan and behavioural simulation is being done. This is in some sense a merging of the ideas of Ad-Hoc and structured DFT, although it is not yet clear what tools are needed for this approach.

5.2.3.1 *Problems of scan design for regular logic*
(1) Burden of area-overhead
Some of these forms of logic are combinational and could, in theory at least, be designed using structured DFT. However, for the sequential logic such as RAMs, systolic arrays, and strip logic, the necessity to convert every sequential element into a scan path flip-flop is overly burdensome.

(2) Problems of fault models and ATG techniques
In many VLSI devices the floorplan is dominated by large regular structures, this implies that the major yield limiting problems will be in these areas. To achieve a high quality test it is necessary to check these structures against a accurate fault model. However, common fault models and good ATG tools for these structures are in short supply.

(3) Burden of long test times
It has been found in practice that for random logic designed to scan path rules the test sequence is limited to at most a few hundred loads of the shift register, but with a good merging method this can be reduced to less than fifty. Hence, test time is rarely a problem. However, arrays such as RAMs where long test times to detect pattern sensitivities may be necessary or ROMs where an exhaustive test is required, the test sequence could be tens or hundreds of thousands of patterns. This is intolerable with the additional overhead of the serial access imposed by the scan path. For large arrays of this type parallel access by bus is necessary.

(4) Performance overhead
Where high performance is essential the introduction of the delays implied by adding the scan logic to the circuit can be unacceptable.

5.2.4 *Off-line self-test*
All of the above DFT methods have been aimed at solving the problems of test associated with test generation, that is ATG and fault simulation. However, the 'VLSI testing problem' is far broader than these limited issues. Many believe that ATE and systems test costs are vitally important issues that must be addressed by DFT. Simple DFT cannot solve these problems on its own,

and methods of off-line self-test called BIST, built-in self-test have been introduced. The basic strategy behind BIST is to trade-off a small overhead in silicon area, against on-chip facilities to generate and check tests on the circuit itself. In this manner, partially or completely self-testing systems can be built from self-testing chips.

In the main the techniques developed for BIST have centred in three groups of techniques:

(1) Scan design methods
These are BIST methods based on either scan path design methods or the BILBO, built-in logic block observer (6). These methods are quite effective because they break the test problem down to combinational logic testing.

(2) Micro-programmed methods
Where an intelligent controller and bus system are available, very efficient test schemes can be created using the BILBO concept. In this area Motorola have led the way with the Mc6804p2 microcomputer (14)

(3) Organic methods
Sometimes regular structures such as RAMs, and PLAs are incompatible with the above self-test methods, either because of their large size or the need for specialist test pattern sequence. In addition, because such structures are regular and easily synthesised, some form of self-test closely matched to the structure is suggested. These methods can be very effective, however, the dense form of RAMs and ROMs implies that the test logic will always contribute a large area overhead unless the structure is very large.

5.3 Fault modelling (4,5,15)

Fault modelling is simply the process of simulating the effect, in the computer, of some physical fault in a circuit for the purpose of test pattern generation or fault simulation. In general, the only faults which are modelled are those which manifest themselves by affecting the logical behaviour of the circuit, these are categorised as,

Stuck-at-faults. (Classical faults)
Bridging-faults.
Stuck-open-faults.
Pattern-sensitive-faults.

(1) Stuck-at-faults
Any physical defect which causes a gate input or output to appear to be permanently connected to a logic '1' or '0' is referred to as a stuck-at-1 (s-a-1) or stuck-at-0 (s-a-0) fault respectively. For a gate which has 'n' inputs there are n + 1 stuck-at-faults, that is the output s-a-1 or s-a-0 and each input stuck-at

the non-dominant logic value. Some physical defects which may be represented by stuck-at-faults are gate input/outputs shorted to the power or ground line, missing contact-holes to gate inputs etc.

(2) Bridging-faults

Bridging faults are the result of signal lines inadvertently coming in contact due to poor oxide isolation between layers, masking defects, incomplete etching etc. The bridging faults, in TTL are modelled as a wired-OR or wired-AND gate depending upon the polarity of the logic; however, in MOS 'matrix' gate structures bridging faults can also alter the overall function realised by the gate. In CMOS the bridging fault can result in a conflict and the strength of the driving transistors determine the outcome. Furthermore, depending upon the location of the bridging fault it is possible for a section of combinational logic to be converted into an asynchronous sequential circuit, which in some instances can only be detected by changing the order of the test inputs.

(3) Stuck-open-faults

This type of fault is a technology dependent fault which occurs in CMOS circuits and converts, for example, a simple 2-input CMOS NOR gate into a sequential element which can store the previous output value in the output node capacitance of the gate.

This particular fault is the result of either a defect in the series or parallel chain which inhibits the node capacitance, under certain conditions from being charged or discharged. Basic functional tests upon the gate will fail to detect the presence of certain stuck-open faults since the order in which tests are applied would mask the fault condition, for example with reference to Fig. 5.3, if input A is assumed to be stuck-open and the test sequence is A = 0, B = 1, followed by A = 1, B = 0 , the output of the gate would be apparently correct for both input conditions, however the output for A = 1, B = 0 is a stored value from the previous input conditions. To detect this fault the input sequence A =

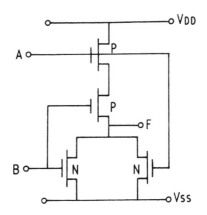

Fig. 5.3   *Two-input CMOS NOR gate*

0, B = 0 would have to be applied before A = 1, B = 0, whereupon it would be observed that the output was a logic 1. Thus test vectors must exercise not only the gross functional behaviour but also the structure of a given logic function, that is in this case all paths which charge or discharge the output node capacitance must be exercised. Furthermore, since the storage effect is capacitive, this type of fault condition is time dependent and may go undetected if dc or quasi-dc testing is used. A model of this fault's storage effect is implemented by a 'gated' latch as shown in Fig. 5.4, the toggle input, T, is controlled by the 'stuck-open fault' network. If there are only classical faults present, then T = 1 and the fault on the gate passes to the output F, however, if a stuck-open fault is present, T = 0 and the output retains its previous value.

(4) Pattern-sensitive-faults

This type of fault condition affects components, for example RAM's, where devices are closely packed and operations performed on one cell affect adjacent cells. The detection of pattern sensitive faults in memory arrays is very time consuming and is made more difficult by the absence of a direct mapping between logical and physical location of the cells requiring the use of a 'Topological Scrambler' in the ATE.

5.4 Practical CAT tools

The development of CAT tools has taken place over the last two decades. Much of the earliest work set very firmly the foundations for later developments. In particular the early work of Roth on the D algorithm and that of Erlich on the concurrent fault simulator are still part of the state of the art in CAT tools. In terms of practically useful tools only three can really be identified in general use:

(1) Fault simulation
(2) Automatic test generation
(3) Testability analysis

5.4.1 Fault simulation

Fault simulation is a necessary adjunct to test pattern generation; it is used, first to determine the fault coverage of a circuit given a set of input test vectors and second to determine what other faults a given test vector will detect, thus reducing the time required for test pattern generation.

In the past fault simulation was affected by modifying a normal or 'true-value' gate level simulator such that gate output values could be held at a logic '1' or '0' simulating a stuck-at-1 or 0 fault, each input test pattern was then applied to determine if the output response differed under faulty and fault free conditions; this process was repeated for every fault in the circuit and became very inefficient when applied to large circuits. In order to improve fault simulation efficiency, three techniques have subsequently evolved, namely,

Parallel fault simulation (16)
Deductive fault simulation (17)
Concurrent fault simulation (18)

5.4.1.1 Parallel fault simulation

In general, within the computer a complete word is used to store the logic value on a node in the circuit. However the logic value can be described, adequately, by using either one or two bits, depending upon the number of values used to describe logic levels in the circuit. Thus, assuming a two-valued logic system, an N-bit word could be used to store the logic value on a node for N independent conditions existing on the circuit. This is the principle underlying the technique of parallel fault simulation, where N copies of a circuit are simulated in parallel, one fault free copy and N-1 faulty copies. Each bit position is assigned to a particular fault condition, thus to determine the effect that a given fault has at any node in the circuit, the appropriate bit position in the word used to store the logic conditions at the given node is examined. The ability to simulate N different conditions in a circuit in parallel is the result of computer instructions operating on words rather than bits. The number of passes required by the fault simulator is the smallest integer greater than $W/(N - 1)$. The fault-free circuit must always be simulated so that the fault conditions detected by a given input pattern can be identified.

5.4.1.2 Deductive fault simulation

In this technique for fault simulation only the fault free circuit is simulated and from the logic values existing at given node the faulty behaviour of the circuit is deduced, either as a result of some fault elsewhere in the circuit or at that node, in this way a list of faults, which may be observed at each node in the circuit, is created. These fault lists are subsequently propagated through the nodes in the circuit until a primary output of the circuit is reached. The contents of the fault lists are continually changing as they propagate through the circuit, since at each gate output a new fault list is computed from the input fault lists using set operations defined by the gate function and the true values on the inputs; certain faults are thus removed from the fault list when they produce the same effect on a node as the fault free signal value.

In deductive fault simulation all faults detected by a given test pattern are identified after one pass through the simulator. The time taken to perform a single pass through the deductive fault simulator is much longer than a single pass through the parallel simulator; since the deduction process is much more complex than simulation, although much shorter than the $W/N - 1$ passes required by the parallel fault simulator to process all faults for a given input pattern. The deductive fault simulator however requires a vast amount of dynamically allocated memory to store the continually changing fault lists.

5.4.1.3 Concurrent fault simulation

In this technique all the faults considered are processed in one pass through the simulator. Unlike the deductive technique explicit gate simulation is used rather than set operations to determine which fault conditions are propagated through the circuit. Again faults which produce the same effect as the fault free circuit are deleted from the fault lists.

The concurrent technique of fault simulation is becoming the most widely used technique of fault simulation since the deductive technique is quite difficult to implement; the parallel technique, although easy to implement becomes very inefficient when used on complex circuits with a large number of faults.

As an example of the concurrent method of fault simulation, consider the circuit shown in Fig. 5.5, the input test pattern is a = 0, b = e = g = 1.

The postulated faults on gate 1 are b/0 (input b s-a-0) and c/1 (output c s-a-1). Since these faults produce an output which differs from the fault free response, they are appended to a fault list and propagated to gate 2.

Postulated faults on gate 2 are a/1, c/1 and d/0, the effect of the fault condition b/0 is also simulated. All of these fault conditions again produce an output response which differs from the fault free circuit.

On gate 3, the postulated faults are b/0, e/0 and f/0. It is seen that b/0 and c/0 produce a response which is the same as the fault free circuit and are thus removed from the fault list.

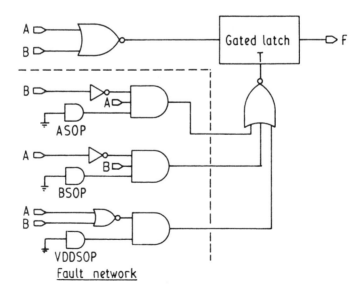

Fig. 5.4   *CMOS fault model*

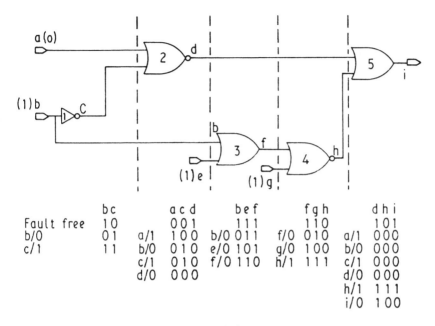

The following data appears beneath the figure:

	bc		acd		bef		fgh		dhi
Fault free	1 0		0 0 1		1 1 1		1 1 0		1 0 1
b/0	0 1	a/1	1 0 0	b/0	0 1 1	f/0	0 1 0	a/1	0 0 0
c/1	1 1	b/0	0 1 0	e/0	1 0 1	g/0	1 0 0	b/0	0 0 0
		c/1	0 1 0	f/0	1 1 0	h/1	1 1 1	c/1	0 0 0
		d/0	0 0 0					d/0	0 0 0
								h/1	1 1 1
								i/0	1 0 0

Fig. 5.5   *Example of concurrent fault simulation*

Postulated results on gate 4 and f/0, g/0 and h/1. Faults f/0 and g/0 are removed from the fault list.

On gate 5, the postulated faults are d/0, h/1 and i/0, the effects of the faults a/1, b/0 and c/1 are also simulated. The fault h/1 is similar to the fault free response, hence the given input pattern would detect the faults a/1, b/0, c/1, d/0 and i/0.

5.4.2 *Test pattern generation techniques*
Although there are many techniques for test pattern generation these can be broadly classified as either Algebraic/Functional, which derive tests from the circuit equations, or Structural which derive tests from topological gate level descriptions.

5.4.2.1 *Algebraic/Functional methods (19):* The most widely used Algebraic/Functional technique is the Method of Boolean Differences, where two expressions are derived for the function of the circuit under faulty and fault free conditions, these expressions are subsequently exclusive-ORed and if the outcome of the operation is a logical 1 a fault exists in the circuit.

In more formal terms, if a combinational circuit realises some function $F(X)$, where X represents a number of Boolean variable $(x_1, x_2, x_3, \ldots x_n)$. In order to observe a fault in one of the variables x_n, the complement of $F(X)$ must be formed when the variable x is converted, i.e. $F(x_1, x_2, x_3, \ldots 1 \ldots x_n)$ EX-OR

$F(x_1, x_2, x_3, \ldots 0 \ldots x_n) = 1$. The expression to the left hand side of the equation is called the Boolean Difference and is represented by the derivative operator d/dx_i and represents all the conditions for which the output function is dependent on the value x_i. To test from a fault on x_i, say x_i stuck-at-1, the logical value opposite to the fault condition is assigned at x_i, that is $\bar{x}_i$ and the tests to detect this fault are derived from the expression $\bar{x}_i.dF(X_i)/dx_i$; the Boolean Difference expression ensures that the circuit function is sensitive to the logic condition on x_i. The main advantage of the Boolean Difference Method is it generates all tests for a given fault, its major disadvantage however, is that if the Boolean function of the circuit is not known, the derivation of the function can be difficult in circuits where the gates have a large fan-in and there is a large number of levels of logic. The Boolean Difference Method may also be used to generate tests for internal faults in the circuit, but this requires a knowledge of the structure of the circuit.

5.4.2.2 Structural methods (20): Structural methods of test pattern generation invariably imply the use of path Sensitisation Techniques where the underlying principle is to propagate information about the possible existence of a fault on a gate internal to a circuit, where there is no direct access, to some observable output node.

The essential steps in the path sensitisation method are first, assume that the inputs to the faulty gate are accessible and assign logic values to these inputs so that the output of the gate becomes sensitive to the fault condition, that is if the fault condition is that the output of a 3-input AND is s-a-0, each input would be assigned a logic 1. Thereafter, information about the logic state on the output of the gate is propagated to an observable output. Propagation of the fault information is achieved by selecting one gate in the fanout list of the faulty gate and making the output of this gate dependent upon the logic input from the faulty gate, by assigning all other inputs to this gate non-dominant values, that is if the gate is an OR gate all inputs not propagating the fault information would be assigned logic 0's. The fault information can now be propagated through one level of logic from the site of the fault, further propagation of the fault information is achieved by selecting a gate from the fanout list of the 'fault sensitive' gate and propagating the fault information through this gate by assigning non-dominant logic values to all the inputs to the gate except the one propagating the fault information: this process is repeated until an observable output node is reached, in this way a 'sensitive path' is set up from the site of the fault to an observable output; this part of the process has a similar effect to generating the Boolean Difference Equation, that is sensitising the output function of a circuit to the logic value on a particular node in the circuit. The final step in the process is to justify the assignment of all the logic values in the circuit to make the gate at the site of the fault sensitive to the fault condition and also to set up the fault sensitive path to an observable output. The justification process is performed using the technique of 'Backward' simulation, which determines the necessary input signals to a gate in order to obtain a given output. The justification process is started with the gate closest to the output of the circuit and is continued backwards until the inputs to the circuit are reached, the values assigned to these inputs are the test for the fault condition.

The above technique only sensitises a single path from the site of the fault to an observable output and would fail to generate a test for certain faults in a circuit which contain a reconvergent fanout structure, this deficiency in the technique is overcome in the D-Algorithm method which sensitises all paths from the site of the fault to an observable output. The essential steps in the D-Algorithm are outlined below.

(*a*) Select a fault condition.

(*b*) Generate the primitive D-cube of failure (PDCF) for the fault condition. The PDCF simply defines the minimal assignment of logic values to the inputs of a gate in order to make the output sensitive to the fault condition.

(*c*) Sensitise at least one path from the site of the fault to an observable output. This is called the D-drive process in which the effect of the fault is propagated through gates until a circuit output is reached. In order to propagate fault information through a gate, the Propagation D-cubes (PDC) for the gate must be derived, these simply define the necessary conditions to be applied to the inputs of the gates, other than those propagating fault information, so that the output of the gate is sensitive to the fault information. The PDC also defines whether or not the fault information will be inverted on passing through the gate. The mechanism for propagating the fault information is called the D-intersection process and comprises matching the logic assignments which already exist in the circuit as a result of the D-Algorithm process with those required to propagate the fault information through a successor gate. The initial D-intersection is performed between the PDCF and the PDC of each gate in the fanout list of the faulty gate, thereafter using the D-intersection process, for each gate through which the fault information has been propagated, an attempt is made to propagate the fault information through the gates in their fanout list. This process is repeated until an output of the circuit is reached.

(*d*) The final stage of the D-Algorithm process is called the Consistency Operation where, using the technique of 'Backward' simulation, the assignment of logic values to the various gate inputs involved in the formation of the Primitive D-cube of Failure and Propagation D-cubes, is justified. The consistency operation starts at the gate output where the fault information can be observed and precedes back to the primary inputs to the circuit. The assignment of logic values to these inputs is a test for the fault condition.

The D-Algorithm was, until recently, the most widely used test generation technique; however, with the increase in complexity of computer systems and the subsequent need for improved reliability through the use of error correction/detection circuitry the D-Algorithm has been found to be inefficient in generating tests for faults in these types of circuits. The characteristic of these circuits which make the D-Algorithm inefficient is the use of exclusive-OR gates with reconvergent fanouts. Consequently an algorithm called PODEM (Path Orientated Decision Making Algorithm) (21) has been developed.

The PODEM algorithm is also based on path sensitisation techniques and the essential steps in the process are summarised below.

(*a*) Set all nodes in the circuit to a 'don't-care' value.

(*b*) Select a fault condition.

(*c*) At the site of the fault, define the necessary gate input values to make the gate output sensitive to the fault condition.

(*d*) Define the 'objective' of setting one of these inputs to the required value.

(*e*) Transfer the objective of setting the input to some value to that of setting the output of the predecessor gate to the same value.

(*f*) Define the necessary inputs to the predecessor gate to achieve the required output value.

(*g*) Define the objective of setting one of the inputs to the predecessor gate to the required value and subsequently transfer this objective to the output of its predecessor gate.

(*h*) Repeat steps (*f*) and (*g*) (back trace procedure) until a primary input to the circuit is reached.

(*i*) Assign the primary input the required value and then use a simulator to determine the effect of the assignment on the circuit. If the initial objective on the gate at the site of the fault has been satisfied, attempt to propagate the fault information to an observable output, if not repeat steps (*d*)–(*i*) until the objective is achieved.

(*j*) Attempt to propagate fault information from the site of the fault to an observable output, again using the backtrace procedure and simulation, as previously described.

Once the final objective of propagating the fault information to an observable output has been achieved the set of assigned primary input values define the test for the fault condition.

5.4.3 Test generation methods for VLSI circuits (21)

The continual growth in the complexity of VLSI circuits has necessitated the development of more efficient test generation schemes. One such scheme developed by IBM Corporation is called PODEM-X. This scheme has been developed around IBM's design for testability philosophy called LSSD; it has been reported (18) that the test generation system has been used successfully on unpartitioned logic modules comprising 50,000 gates. The system comprises three test pattern generation programs, a fault simulator and a test compaction program.

The first test generation program called SRTG (Shift Register Test Generator) is used to perform functional tests on the blocks of shift registers which are used in the LSSD method to partition the system into less complex functions and which are used, during the testing phase, to apply input test patterns to the partitioned blocks and also to store the output responses.

The testing strategy employed in PODEM-X is to generate, initially a set of 'global' tests which are designed to detect a large number of faults, thereafter to generate a set of 'clean-up' tests which are designed to detect the faults which are not covered by the 'global' tests. The global test generator is called RAPS (Random Algorithm Path Sensitisation) whose objective is to derive input patterns which will sensitise a large number of random paths through a circuit and thereafter to use a fault simulator to determine which faults are detected by the patterns. Specific tests are then generated for the remaining faults by the test generation procedure called PODEM which has been described previously.

The fault simulator in PODEM-X is called FFSIM (Fast Fault Simulator) and is used to determine the faults detected by tests generated using the RAPS and PODEM procedures.

Invariably, when generating tests for complex circuits not all inputs will have assigned values, this can be used to good advantage since it permits test patterns to be merged together thus reducing the overall number of tests to be applied to the circuit. This merging or compaction process can either be performed statically or dynamically. In static compaction the tests are merged after the test generation procedures have been completed; whilst in dynamic compaction the process is performed during the test generation phase.

5.4.4 Testability aids

Previously various approaches to testability have been discussed. Some of these approaches enforce testability by rule, hence to ensure testability it is necessary to check the circuit against the rules. For circuits of VLSI complexity even this can be a difficult and time consuming task. Other approaches are less strict and only make recommendations, with this approach rule checking is of little value. For the Ad-Hoc approach a number of testability analysis tools have been developed. These tools don't check a circuit for rule violations but attempt to assess testability by the yardstick of controllability and observability. Using these terms the testability of circuits can be assessed.

5.4.4.1 Testability analysis: During the seventies and early eighties testability analysis was a very popular idea within the VLSI testing community. However, as has been previously indicated doubts have been cast on the value of these tools, nevertheless, they may still have some value in specific instances. Especially, for small gate array design or PCBs where testability can be a problem but structured DFT is impractical.

Two testability tools are widely available, SCOAP and Gen-Rad's HI-TAP program, previously known as Camelot; these two tools are discussed below (10,11).

(1) SCOAP

SCOAP can be usefully applied to the development of easily testable circuits as part of a scheme to improve designer awareness of the problems of testing and the solutions available. In this manner, knowledge of the testing problem can be cumulatively developed by individual designers and by the design group without the painful and expensive learning cycle associated with the direct use of test generators.

The method calculates six functions to characterise the controllability/observability (C/O) properties of the circuit. From these functions a quantitative measure of the cost of controlling and observing the internal nodes of the circuit can be derived. This is achieved solely by considering the topology of the circuit without reference to a specific set of test vectors or test-generation method.

SCOAP analysis assigns six testability values to each circuit node: combinational controllability 0 and 1 (CC0, CC1), sequential controllability 0 and 1 (SC0, SC1), combinational and sequential observability (C0, S0).

These are defined as follows: the CC values are defined as the minimum number of nodes that must be set in order to produce either 0 or 1 on the node in question. The C0 value of the node defines the number of nodes that must be set in order to propagate the value of that node to the primary output plus the number of combinational cells on the sensitive path. Hence, the combinational testability values give an upper-band estimate of the cost of the test generation in a spatial sense (i.e. the proportion of the circuit that must be instantaneously controlled in order to test a node). Sequential testability values operate in a similar manner, except that they refer only to the sequential nodes of the circuit (i.e. memory elements). Thus sequential testability gives an upper-bound estimate of the cost of test generation in a temporal sense. In this manner the difficulties of testing a particular part of a circuit can be isolated to specific spatial and temporal characteristics, and the appropriate measures can be taken to improve testability.

A simple example of the application of SCOAP is shown in Fig. 5.6.

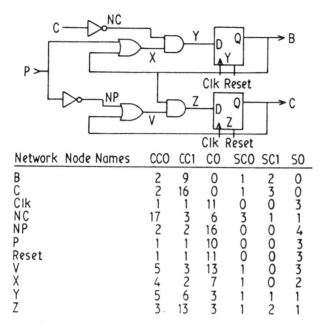

Network Node Names	CCO	CC1	CO	SCO	SC1	SO
B	2	9	0	1	2	0
C	2	16	0	1	3	0
Clk	1	1	11	0	0	3
NC	17	3	6	3	1	1
NP	2	2	16	0	0	4
P	1	1	10	0	0	3
Reset	1	1	11	0	0	3
V	5	3	13	1	0	3
X	4	2	7	1	0	2
Y	5	6	3	1	1	1
Z	3	13	3	1	2	1

Fig. 5.6 *A simple SCOAP example*

A weakness of SCOAP in its present form is that it is applied too late in the design cycle of the project to be totally effective. Ideally testability analysis in some form should be available at the highest levels of design to ensure that design trade-offs at this level have no adverse effects on the final testability of a design.

In SCOAP controllability is measured in four values at each node, CC0, CC1, SC0, SC1 these are combinational controllability zero/one and sequential controllability zero/one. These are measured in integer values between 1 and 999999, the value represents the number of nodes that need to be controlled within the circuit to achieve that aim. Hence, a primary input has CC0 and CC1 set to 1 and SC0 and SC1 set to 0. This is because the analysis distinguishes between combinational nodes, all the circuit nodes and sequential nodes, all nodes driven by a sequential logic function. Thus an input need only be controlled itself to be set to zero or one. The controllability values for the whole circuit can then be calculated by proceeding from inputs to outputs calculating the minimum number of nodes needed to set each output to a particular state. Once the controllability values are known it is then possible to calculate the two observability values, CO, Combinational Observability and SC, Sequential Observability. This is achieved by backtracking through the circuit from outputs to inputs calculating the minimum number of nodes that must be set, both combinational and sequential, to observe that input via an output.

One feature of the SCOAP analysis is that because it initialises all nodes initially to worst case, it is very good at detecting initialisation difficulties that can cause major testability problems.

(2) HI-TAP/Camelot

In Camelot controllability, denoted by CY, is constrained to be in the range 0 to 1. The maximum value of 1 represents a node, such as a primary input, where it is easy to assign a 1 or 0. The converse of course is a node such as a floating pin, this is uncontrollable and is assigned the value 0. The observability, OY, is similarly 1 for a primary output and 0 for the unattached output pin. Given that the CYs are defined for the inputs it is possible to calculate the CYs of the internal nodes using the CTFs or controllability transfer factors. This is done by proceeding from the inputs through each gate to the primary outputs. The CTF of a gate or function is closely related to the number of zeros and ones in the output truth table. Given the CYs for each node, the OYs can then be calculated. This is done by back-tracking from the primary output to the inputs using the OTF values. The OTF is the observability transfer factor, these are based on the number of sensitive paths that can be created between inputs and outputs. This is of course related to the CY values at the inputs to each gate or function. Compared to SCOAP Camelot is rather crude in that it fails to distinguish between zero and one controllability and has no concept of combinational/sequential testability. It has the advantage of being easy to understand, and clearly focuses attention on problem areas of the circuit.

5.4.4.2 Testability rule checking: Where structured design for testability methods are used rule checkers of some type are almost essential. It is only too likely that designers either in error or maliciously design parts of a circuit such that it breaks some basic rules and renders the circuit untestable. However, the problem here is not checking rules but formulating sensible rules. It is quite easy to formulate testability rules that ensure a testable design, but at the same time are impractical. At the gate-level or *structure-level*, it is easy to formulate a closed set of design rules such as LSSD or the scan path rules, however, not all designs or designers can comply with such rules and any that cannot will find their test needs unsupported by CAD tools. However, these design rules are usually quite easy to check, i.e. check for a scan path with controllable clock signals and for purely combinational logic in the remainder of the circuit (22). It has been found in practice that DFT rules that specify the *behaviour* of a useful scan path or data bus, and not its structure are more acceptable to designers and encompass a wide range of circuits. Thus a more liberal DFT scheme can be created by using behavioural DFT rule, in addition, because the behaviour of a circuit is known in advance of the structure this means that DFT rules are checked much earlier. Of course, it is still necessary to check that the structure is testable, but this is trivial where 'Design is correct by construction' as in a silicon compiler or Macro-cell assembler, whilst in hand-crafted designs it is the designer's responsibility.

5.5 Automatic test equipment

Current microelectronic circuits are challenging the capabilities of available automatic test equipment (ATE) to test them. The new heights of complexity, speed, and the variety of functions implemented on a single chip require that the performance of the next generation of ATE be significantly better than the present. Complexity of the circuits, large pin counts and the mixing of analogue and digital sub-systems are currently the most taxing problems and even with new high-performance testers long test times are the norm. A common problem is the introduction of both logic and large memories on a single chip and hybrid analogue/digital circuits that demands the use of general purpose ATE in place of the specialist test systems currently used.

High-performance and, in particular, high-frequency circuits bring specific test problems. Wafer testing (i.e. testing before packaging) with other than superficial functional checks is often impracticable, and parametric tests are conducted on the packaged device. No single ready-made solution to the testing of this class of circuit exists.

Past developments in ATE that have done much to reduce testing costs are the use of computer controlled systems that can interface to the IEEE 488/IEC 625 bus and the concept of 'virtual instrumentation'. In the main, these have only been used in dedicated systems for special analogue/digital IC functions. In the future the concept of a generic tester architecture with a family of machines offering a series of cost/performance options is required. At the lower end a minimal cost tester for debug/design verification should exist, this would naturally interface to a workstation network. The low-end machine of the series should be able to emulate the functional features of more expensive machine using software. In the higher series machines most of the features would be implemented in hardware to meet the performance required of a production test system. Only in this way would it be possible for designers to achieve the fullest utilisation of the tester architecture with a particular IC test program, yet not waste valuable tester time. It must always be remembered that ATE is the single most expensive LSI production kit to both purchase and run.

5.6 Conclusions

Although sophisticated test generation schemes have been developed to overcome the problem of testing VLSI circuits, these only attack the symptom of the problem, namely complexity, and hence can only offer an interim solution to the problem. Alternative approaches, for example Design Testability Methods, attack the root cause of the testing problem, which is the controllability and observability of signal nodes in the circuit, and will offer a more long term solution to the testing problem. For most of the industry to really solve the testing problem it is necessary to consider the whole life-cycle testing problem of a VLSI system. If this is not done the economics of testing cannot be optimised. Care must be exercised, however, as the wrong cure can be as fatal as the disease in testing.

References

1 EICHELBERGER, E. B. and LINDBLOOM, E.: 'Trends in VLSI testing', VLSI '83, *Elsevier Science Publishers*, 1983, 339–348

2 GRASON, J. and NAGLE, A. W.: 'Digital test generation and design for testability', *17th Design Automation Conference Proceedings*, June 1980

3 MANGIR, T. E. and AVIZIENIS: 'Failure modes for VLSI and their effect on chip design', *IEEE Proceedings 1st International Conference of Circuits and Computers*, May 1980, 685–688

4 BREUER, M. A. and FRIEDMAN, A. D.: 'Diagnosis and reliable design of digital systems', *Computer Science Press*, 1976, USA

5 MUEHLDORF, E. E. and SAYKAR, A. D.: 'LSI logic testing — an overview', *IEEE Trans. Computers*, C-30(1), January 1981, 1–17

6 LALA, P. K., 1985: 'Fault tolerant and fault testable hardware design', 1985, Prentice-Hall, London

7 WILLIAMS, T. W. and BROWN, N. C.: 'Defect-level as a function of fault coverage', *IEEE Trans. on Computers*, C-30(12), December 1981, 987–988

8 WILLIAMS, T. W. and PARKER, K. P.: 'Design for testability — a survey', *IEEE Trans. on Computers*, C-31(1), January 1982, 2–15

9 BENNETTS, R. G. and SCOTT, R. V.: 'Recent developments in the theory and practice of testable logic', *Computer*, 9(6), June 1976, 47–62

10 GOLDSTEIN, L. H.: 'Controllability/observability analysis of digital systems', *IEEE Trans. Circ. and Syst.*, CAS-26, September 1979, 685–693

11 BENNETTS, R. G., MAUNDER, C. M. and ROBINSON, G. D.: 'Camelot: A computer-aided measure for logic testability', *Proc. IEE, 128* Pt.E, September 1981, 177–198

12 AGRAWAL, D. A. and MERCER, M. R.: 'Testability mesures — what do they tell us?', *IEEE Inter. Test Conf*, 1982, 391–396

13 GRIERSON, J. R., et al.: 'The UK5000', *Proc. 20th Design Automation Conf.*, 1983, 629–636

14 KUBAN, J. and BRUCE, B.: 'The MC6804p2 built-in self-test', *Proc. IEEE Inter. Test Conf.*, 1983, 295–300

15 WADSACK, R. L.: 'Fault modelling and logic simulation of CMOS and MOS, integrated circuits', *BSTJ, 57(5)*, May–June 1978

16 BENNETTS, R. G.: 'Introduction to board testing', *Computer Systems Engineering Series*, Edward Arnold, 1982

17 ARMSTRONG, D. B.: 'A deductive method for simulating faults in logic circuits', *IEEE Trans. of Computers*, C-21(5), May 1972

18 BOTTOROFF, P. S.: 'Computer aids to testing — an overview", *NATO Advanced Study Course*, Urbino, July 1980

19 SELLERS, F. F., HSIAO, M. Y. and BERNSON, L. W.: 'Analysing errors with Boolean differences', *IEEE Trans. on Computers, C-17(7)*, July 1968

20 ROTH, J. P., BOURICIUS, W. G. and SCHNEIDER, P. R.: 'Programmed algorithms to compute tests to detect and distinguish between failures in logic circuits', *IEEE Trans. on Computers, EC-16(10)*, 1967, 567–580

21 GOEL, P. and ROSALES, B. C.: 'PODEM-X: An automatic test generation system for VLSI logic structures', *18th Design Automation Conference Proc.*, June 1981, 260–268

22 BHAVSAR, D. K.: 'DFT calculus: An algorithm for DFT rules checking', *Proc. 20th D.A. Conf.*, 1983, 300–307

Chapter 6

Auto Layout

D. J. Kinniment
University of Newcastle upon Tyne

6.1 Introduction

Automation of the layout of an integrated circuit is usually divided into a series of sequential tasks:

(1) Partitioning of a very large system into smaller subunits which may be realised as chips, or be part of a single large chip.

(2) Placement of the subunits on a chip into absolute or relative locations to minimise the overall area and ensure that the final stage of finding the interconnections is possible.

(3) Routing of the actual interconnections.

The reason for this division is to reduce a very difficult problem into a set of simpler tasks, but there are design tools which combine placement and routing, or other parts of the problem, and there are good reasons for doing so provided the computational task does not become intractable as a result.

Partitioning a VLSI system during the design phase is no different in essentials from designing any other large system since in taking it from the specification to realisation, the design must be divided down into the subunits, and these further subdivided until known realisable objects are reached. This decomposition process is best done from the top down, starting from a clear statement of the required function of the final product.

Computer aids to the layout process, however, may operate in a top down or bottom up fashion and the characteristics of the results achieved by these layout methods are different.

6.1.1 Abutment

One methodology is typified by the INMOS and ASTRA systems (1,2), in which the whole area of the chip is divided into small areas intended to contain the major components of the system, with interconnection points at the boundaries of these component areas to define the signal flow. This establishes the main floor plan, and further subdivisions of the major components establishes a floor plan within each of the components.

The advantages of this approach are that connections between components cells always lie on adjacent boundaries, and the area occupied by connections is therefore minimised. The physical mapping done by the layout follows the design decomposition process, and area efficient chips of good performance can be produced. One of its disadvantages is that accurate area decomposition at the top level of the hierarchy depends on a knowledge of the likely component sizes, and those are not usually known until the bottom level physical detail has been completed. Much skill is therefore required to reduce the number of iterations up and down the design hierarchy during layout. Another problem is that at the bottom level in the hierarchy, the leaf cells often have to be deformed, or the pin positions altered to allow connections to a neighbour by simple abutment, and consequently much detailed design has to be done.

6.1.2 Place and route

In a VLSI chip, the problems of layout lie more in the immensity of the design task, than in the need to achieve minimum area, so that techniques which assume the physical interface of a component (the outline, area, and the pin connection points) is fixed are preferred. If these characteristics are known, or can be computed, for a library of component cells, then a bottom up approach can provide a simple, quick, and effective layout technique. A set of primitive, or leaf cells grouped together at the bottom level of the hierarchy can be placed close to each other, and connections made by routing wires around or through the cells so that they make up a larger component cell whose outline, area, and the connection points of this are then known accurately. Only one pass through the layout process is then necessary for a correct design, though the silicon area may be larger than optimum.

Chip design methodologies which rely on a bottom up, place and route strategy are often referred to by the term 'semi custom design'.

6.1.3 Chip architectures

The design tools used to implement place and route methods are significantly affected by the chip architecture on to which the design is to be physically mapped. The main difference between these architectures lies in the fixed or flexible nature of the areas available for routing wires.

A large class of chip architectures are characterised by gate arrays, in which fixed primitive cells are prediffused into the silicon, usually with wiring channels in between rows of cells. The cost of production of new designs using such a fixed format is reduced because new mask sets only have to be produced for the variable interconnection layers, but the penalty is that it may not be possible to accommodate all the connections required for a particular design in the channel space provided. Some work has been done based on empirical considerations to estimate the space required for a gate array (3) but no guarantee can be given that any design will necessarily be routable at present.

Other methods assume that it will be possible to move cells apart sufficiently to accommodate the connections, and pay the cost penalty of customising all the masks for every design.

6.2 Placement methods

The main aim of the placement algorithm in gate array and other semicustom structures is to provide the basis for the following routing process.

The computation required to investigate all possible placements of component cells on a chip is prohibitive because the number of different placements is proportional to n! where n is the number of components to be placed. Many methods attempt to estimate the 'goodness' of partial placements as the process progresses, choosing only the 'best' module to place next in the 'best' location. This involves continually estimating the 'cost' of selecting a module and placing it in a particular location. The process of cost estimation must be fast in order to explore as many placement variations as possible, but must also bear as close a relation to the actual routing process as possible to be realistic.

Typically placement programs construct an initial placement by one of two methods, either a cluster development technique on the lines described above, or one based on the idea of attractive and repulsive forces between components. If all components to be placed are considered to be joined by elastic connections, then the attractive forces between modules are proportional to the number of connections and the distance between the points joined. Repulsive forces are provided by the proximity of the modules themselves. Solution of the equations of motion gives a rough placement that is essentially 'force directed' (6).

After initial placement, considerable improvement can be achieved by interchanging components within local groups. The effect of interchanging within groups of up to ε components in terms of the amount of computation required to achieve a given reduction in total wire length is shown in Fig. 6.1, which is derived from work by Goto (4).

Estimates of goodness by wire length provide a guide to placement improvement, unfortunately the improvement process tends to bunch components in a relatively small area, leading to difficulties of congestion in the wiring channels in that area, and Figs 6.2 and 6.3 show the connection

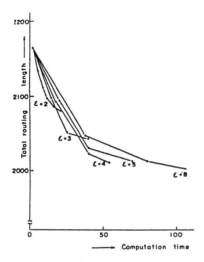

Fig. 6.1 *Computation and wire length*

paths found for a gate array design placed manually in Fig. 6.2, and placed by an algorithm which attempts to minimise wire length in Fig. 6.3. These clearly indicate a build up of components in the top right hand corner of the automatically placed chip, which caused failure of connections to route (the numbered squares indicate the failed connection points).

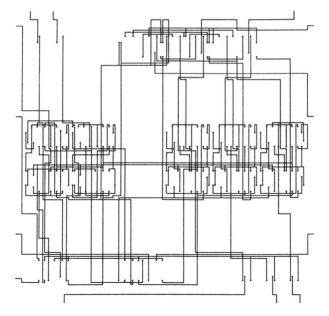

Fig. 6.2 *Manual layout*

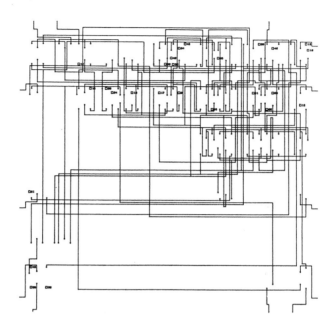

Fig. 6.3   *Congestion due to wire length minimisation*

It is clearly better to attempt to minimise the congestion in the wiring channels rather than the wire length itself, and many systems now do this, using weighting functions which depend on the probability of a wire using a particular channel and the likely occupancy of that channel by the other wires.

Rather than simply placing pre-existing modules, some methods attempt to partition the network in a more rational way, for example, the Mincut Algorithm (5). Modules are placed to the left or right of a cut line in such a way as to reduce the number of connections crossing the line to a minimum. The partitions are themselves cut, and so on. The process is illustrated in Fig. 6.4 in which a graphical representation of the design shows the modules as arcs in the graph and the channel between as nodes. Since the partitioning reduces the number of connections in a channel, routing problems should be eased.

There are many other initial placement methods, but whilst adequate results are now being produced they are still inferior to the human designer in some situations and particularly where there is an easily identifiable structure to the design.

Placement improvement can also be carried out on the basis of a purely random selection of cell interchanges to effect improvement, but if only the interchanges leading to improvements in the cost measure are retained, it is possible that a local minimum will be obtained rather than a global minimum. Attempts to overcome this deficiency make use of the idea of 'Simulated

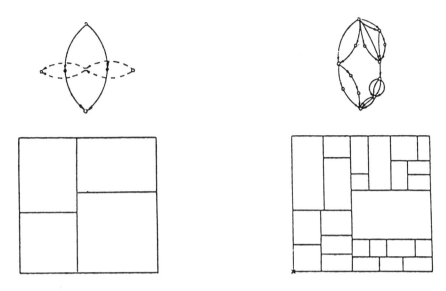

Fig. 6.4 *Mincut procedure*

Annealing' in which the probability of accepting an interchange is a function a) the cost, and b) how many changes have been made to the initial placement. At first, changes which lead to quite large increases in cost may be accepted, but as the process proceeds the probability decreases.

6.3 Routing

After placement of components, the routing software must determine the precise paths of the interconnections between connection points. Again this is a problem in which the very large number of connections required, interact with each other in a way which makes the exploration of all possible paths impractical for chips of present complexity, and algorithms have to be found which can provide a solution in an acceptable time without overflow tracks for gate arrays, or within an acceptable area for cell based chips.

The problem of routing can itself be split up into simpler problems and in one possible subdivision the connection paths for each set of connected points are found in turn. The sets of points which are to be electrically connected are termed nets, and the first step is to form a list of nets ordered in a way intended to make the overall routing process likely to be successful.

Each net is then taken in turn and the detailed paths between the connection points found.

At the simplest level there are many algorithms and heuristics which have been used to join a net of points. A detailed description of all these methods and their relative merits is beyond the scope of this chapter, and only the essential points of a few will be outlined.

6.3.1. The Lee path connection algorithm

Perhaps the most commonly used method for finding the shortest path between two points was described by Lee, in 1961 (7).

In essence this consists of choosing one point as the source, S and the other as the target T. The area to be explored to find the path is divided into a grid of cells each of unit size, and is represented by a data structure such as an array. The source cell in the array is then marked with an integer. The address of this marked cell is then placed into a list called the frontier cell list and all unmarked neighbours of the frontier cells are then marked with the next integer. When no more unmarked neighbours exist, the old frontier cell list is deleted and replaced by the list of newly marked cells. The process is then repeated as shown in Fig. 6.5 until the target cell is hit.

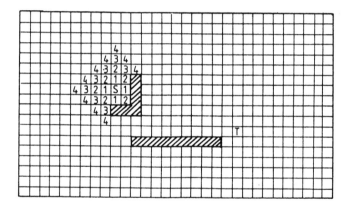

Fig. 6.5   *Lee's algorithm*

If the frontier cell list has no members at any point before the target is reached, there can be no path between the source and target. Lee's algorithm thus guarantees that if a path exists it will be found, and that the path found will be the shortest since at any time the minimum path length to every frontier cell is equal.

The actual path can easily be retraced by following back from the target to the source looking at each point for the previous integer in the sequence.

The reason for the ubiquity of Lee's algorithm is its adaptability. It can easily cope with multipoint nets by expanding from all points simultaneously and retracing in both directions from a contact point, and can also be modified to deal with multilayer metal by using a 3 dimensional array structure to represent the grid rather than the 2 dimensional structure of Fig. 6.5.

A modification of Lee's algorithm has been described by Rubin (8), in which the cost of expanding a particular cell can be made a function of the direction of expansion (less for movement towards the target, more for movement away). Where a large number of possible cells with the same cost could be expanded, Rubin proposed that the latest cell added to the cell list be expanded. Such an algorithm is known as a 'depth first predictor'.

The actual algorithm is as follows: its action is illustrated in Fig. 6.6.

(1) Set a cost threshold to zero.
(2) Find the last cell, 'c', in a cell list whose estimated cost equals the cost threshold. If none, set the cost threshold to the least estimated cost of any cell on the list and repeat step 2.
(3) If the cost threshold exceeds the maximum allowable cost for this track, Exit (the path is too expensive).
(4) If 'c' is a target cell, note its expansion direction in a 'mark' array and Exit.
(5) If 'c' is already marked, go to step 8 (the cell has been expanded).
(6) Add each of the unmarked neighbours of 'c' to a cell list with their cost.
(7) Note the expansion direction of 'c' in the 'mark' array.
(8) Delete 'c' from the cell list.
(9) If the cell list is empty, Exit (no path exists).
(10) Go to step 2.

The cost of a neighbour cell in Fig. 6.6 is the sum of its distance from the prime target and a fixed expansion cost of one unit per move. Since the latest minimum cost cell is always the one expanded, expansion takes place preferentially towards the target.

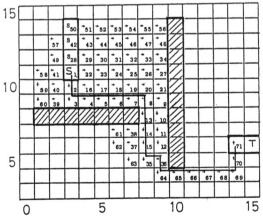

Fig. 6.6 *Expansion towards target*

This algorithm can also be modified to favour the X direction on one layer, and the Y direction on the other, as well as adding penalities for other obstacles, such as vias.

Another problem with grid based routers is the size of the data structure required to hold the complete chip map in terms of an array of small cells. Typically each cell must be able to represent several states, empty, no go, source, target, and 2–3 marking states. One byte per cell might therefore be reserved, and a 10 mm × 10 mm chip with 2 layer metal on an 8μ grid requires $2 \times 1250 \times 1250 \simeq 3$ M bytes. It is easy to see the problems of submicron geometries.

6.3.2 Heuristics

Cellular routers such as Lee's algorithm may explore very large areas of board before a successful connection is found and since the area is explored on a cell by cell basis the computation required will be considerable. If the number of connections is proportional to the number of gates n, and the average length proportional to $n^{\frac{1}{2}}$, the area explored per connection will be at least proportional to n^2 in a simple Lee router and the total layout time at least proportional to n.

Usually most of the computation time is spent on the last few connections since these are often difficult to find and most of the i.c. area has to be explored to find a clear path.

Usually, however, about 75% of connections are relatively simple, involving nothing more than a straight connection, or one with at most one or two bends. A heuristic approach which attempts to find these simple connections quickly by extending parallel lines from the two points and connecting between the lines with an orthogonal line was first described by Aramaki (9), and has the advantage of following an x-y path which as well as being quick to compute is less likely to block subsequent nets than the more complex paths sometimes produced by Lee's algorithm. Fig. 6.7 shows the kind of paths found by a simple heuristic using two layouts of interconnect.

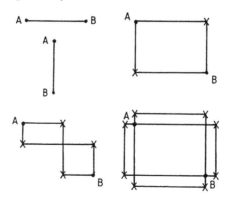

Fig. 6.7   *Possible routes with 2 or fewer bends*

A heuristic like this requires that the net be split into successive pairs of points which are to be joined. This is a relatively easy task and simple algorithms which start with a central point in the net joining further points, until no points remain unconnected, can be quite successful.

6.3.3 Channel routers

Rather than looking at each net in turn and dividing all the area available into a grid of cells, many recent routers examine each row or column in order to pack connections optimally in the wiring channel formed between the rows or columns. An early method was described by Hashimoto and Stevens (10). Here all the nets participating in the use of a particular channel are considered together, and the aim is to minimise the channel width required to provide the connections and through routes. Another advantage is that the connections are stored as wire segments with start and end points only, using less memory than the cells of grid routers, and usually faster in testing whether or not a segment exists at a particular position.

There are many published channel routing algorithms (10,11), one of the simpler being known as the 'left edge' algorithm, which is shown in Fig. 6.8. In a horizontal channel with terminal points placed on the lower and upper edges. The first terminal point found is connected to all the other terminal points of that net by allocating a horizontal connection to the first available track nearest the upper edge. Horizontal connections are routed on one layer and a second layer is assumed to be available for perpendicular connection segments linking to the terminal points. The next unconnected terminal point is then found and the process repeated.

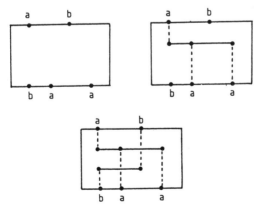

Fig. 6.8   *Channel router*

Not all terminal assignments can be routed with a fixed channel length, for example the situation of Fig. 6.9 cannot be solved without providing extra channel space or using 'dog leg' structures.

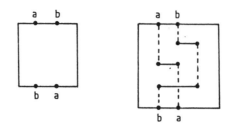

Fig. 6.9   *Net conflicts*

6.3.4 Hierarchical routing

Gate array routers often tackle the whole routing problem in one pass using an algorithm such as Lee's algorithm, but the amount of computation required is at least proportional to the square of the number of gates. Considerable savings in time can be made by the use of local channel routers which consider the detailed wiring of a relatively small area in a big chip without regard to the surrounding area. If each small area is considered in turn, the computation will clearly be not much greater than proportional to the number of gates, but there is a need for an earlier, global planning stage to assign nets to channels before the detailed routing is done, and there may be the possibility of backtracking to the global stage again if a particular channel proves impossible with the initial allocation of wires.

In a global router, the chances of a connection going through a particular channel are estimated, total congestion from all the connections calculated for each channel, and compared with the channel capacity, by using an empirically derived cost function.

The cost of putting the wires through each channel can now be estimated, and the global router tries to decide which channels to actually use in order to minimise the cost.

There are two areas where further improvements in computing time and economy of memory can be made within this scenario: these are

(*a*) The use of prewired macros. The greater the regularity factor in the design the greater advantage can be taken of laying out a block of gates or modules once and copying that block in the positions required, rather than computing in detail every part of the chip area.

(*b*) Greater use of hierarchy. There is no reason to stop at two levels in the hierarchy of the design (global and local), and some advantage may be gained from matching the layout hierarchy to the conceptual hierarchy in the input text provided by the designer. This input should also be aimed at regularity and reduction of the number of connections between sub-systems – both qualities helpful to autolayout – because these aims also reduce the complexity of the design.

A router which combines global and local tactics in a hierarchical strategy is described in detail by Liesenberg (12).

Given a placement in which only the relative positions of the cells in a group are described, it is possible to derive a graph representing the channels, Fig. 6.10. Individual connection points for the terminals on the component cells can then be inserted in the graph as in Fig. 6.11 and the most appropriate path through the graph to join these points is then found.

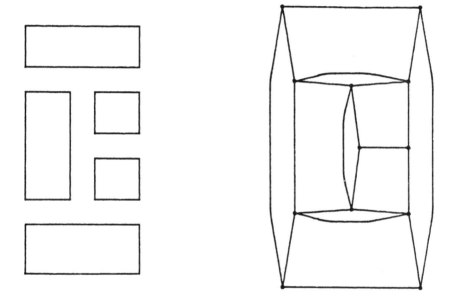

Fig. 6.10   *Relative placement and graph*

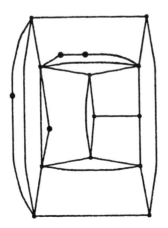

Fig. 6.11   *Routing path through graph*

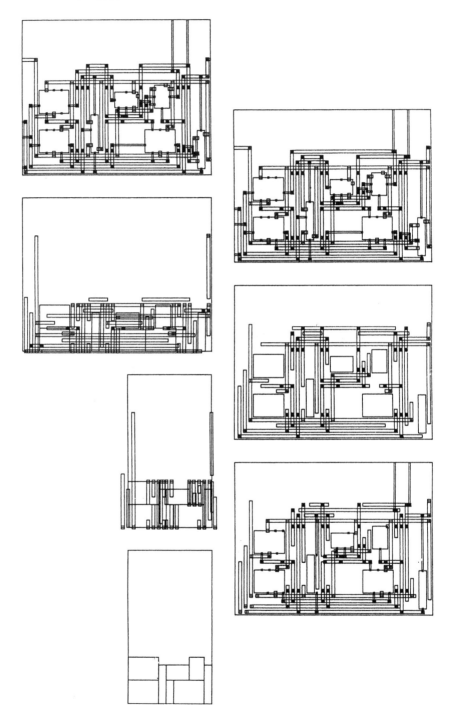

Fig. 6.12 *Layout expansion*

In order to ensure that ground and power connection can be routed on the same layer, usually that with the highest conductivity, ground connections are always placed on the left or top edges of cells and power on the right or bottom edges. A simple algorithm which makes ground connections with segments on the extreme right or bottom of a channel and power connection on the left or top of the channel will ensure that they do not cross.

This global graph searching process assigns connections to channels, and is followed by a detailed channel routing process which finds the actual channel width. Unfortunately, however, the detailed routing of each channel is dependent upon other interacting channels since vertical and horizontal channels intersect. A solution to this problem is to lay out all horizontal channels first, followed by all vertical channels. The individual horizontal channel layouts are thus independent, and a close estimate can be obtained for the necessary width. Given the detailed horizontal channel layout and the corresponding vertical distances, the vertical layout is then attempted. This may, in turn, give rise to changes in the lengths of horizontal channel connection, and possibly their order, but these changes are minor. Iterating successively between horizontal and vertical, ensuring that the channel widths are only increased and not decreased guarantees an eventual solution as shown in Fig. 6.12.

6.4 Conclusions

The job of routing is made much easier if sufficient room is made available for the tracks. This is easy in a full custom design, just allocate sufficient for the particular case.

It is more difficult with gate arrays or cell arrays with fixed placements since providing enough to cover all cases is a waste of silicon in the simpler cases. If insufficient has been provided in a gate array chip, it cannot later be rectified by better CAD, much computation would be done with little return. Recourse must be made to interactive graphics, at best an error prone and difficult process, and particularly difficult after algorithms have been used which guarantee to find a path if it exists since other tracks must be removed to find space.

Typically, programs to connect gate arrays spend most of their time looking for these last few connections, which are usually long, and the whole chip area must be searched to find if a path exists. Often the paths do not exist because of earlier decisions which prevented access to critical points and large amounts of the wiring must be altered to make the required connection. Channel routing based techniques based on placement expansion methods can guarantee 100% routing and may also be considerably faster, particularly if the design hierarchy is exploited. Since the interconnection layers often represent the limit on performance and device packing densities on a chip, it is likely that future developments in process technology will concentrate on increasing the number of layers available, and that layout techniques will change as a result.

6.5 References

1. CHESNEY, M.: 'The INMOS design approach and associated tools', in *VLSI architecture*, 1982, editors B. Randell and P. C. Treleaven, Prentice Hall.
2. REVETT, M. C., and IVEY, P. A.: 'ASTRA – A CAD system to support a structural approach to I.C. design', in *VLSI '83*, 1983, editors F. Anceau and E. J. Aas, North Holland.
3. DONATH, W. E., and MIKHAIL, W. F.: 'Wiring space estimation for rectangular gate arrays', in *VLSI 81*, 1981, editor, J. P. Gray, Academic Press.
4. GOTO, S.: 'An efficient algorithm for the two dimensional placement problem in electrical circuit layout', 1981, IEEE Trans on Circuits and Systems, *CAS 28, No. 1*.
5. LAUTHER, U.: 'A Min-Cut placement algorithm for general cell assemblies based on graph representation', in *ACM IEEE 16th Design Automation Conference*, 1979, San Diego.
6. UEDA, K., and KITAZAWA, H.: 'An algorithm for VLSI chip floor plan', 1983, Electronics Letters, *Vol. 19, No. 3*.
7. LEE, C. Y.: 'An algorithm for path connections, and its applications", 1961, IRE Trans, *EC 10, No. 3*.
8. RUBIN, F.: 'The Lee path connection algorithm', 1974, *IEEE Trans, C-23, No. 9*.
9. ARAMAKI, I., KAWABATA, T., and ARIMOTO, K.: 'Automation of etching pattern layout', 1971, *CACM, No. 14*.
10. HASHIMOTO, A., and STEVENS, J.: 'Wire routing by optimising channel assignment within large apertures', 1971, In 8th Design Automation Workshop.
11. BURSTEIN, M., and PELAVIN, R.: 'Hierarchical channel router', 1983, Integration, The VLSI Journal, *Vol. 1, No. 1*.
12. LIESENBERG, H. K. E.: 'A layout module for a silicon compiler', 1985, Ph.D. thesis, University of Newcastle upon Tyne.

Chapter 7

Symbolic design of VLSI circuits

P. Ivey

7.1 A definition of symbolic design

If we look up a dictionary definition of 'symbolic design' we might get something like this:

Symbol – A letter, figure or sign used to represent a quantity, phenomenon, operation, function etc. . . .

hence *Symbolic Design* is design by means of letters, figures etc.

Design – to work out the structure or form of, as by making a sketch, outline, pattern or plans.

Design, therefore, is the decomposition of behaviour into structure (see Fig. 7.1).

The important thing is that, by this definition, design using symbols can be carried out at all levels of the hierarchy from the highest to the lowest. The aim

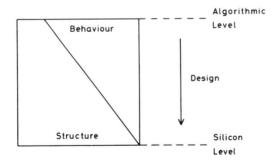

Fig. 7.1   *Design – behaviour to structure*

of the method is to devise symbols that convey just enough information to allow you to make decisions at the level at which you are currently working. However, it is crucial that, in designing your symbols, you do not 'throw the baby out with the bath water' and abstract out vital information and leave a worthless symbol.

Examples of good symbols might be Engineering Drawings, Architects plans etc. Examples of poor symbols might be 'text' in physical design or schematic capture in VLSI design. The good symbols capture just enough information to make them useful while the poor symbols have 'lost' some of the crucial details.

7.2 A short history of symbolic design

Historically the design of the layout of full custom integrated circuits has had two major branches. In the older method, computer graphics systems have been used to assist the manipulation of polygon data. Such a system could equally well be used to design buildings and hence it has no in-built 'knowledge' of integrated circuits. This leaves the problem of the correctness and functional accuracy entirely with the designer. And, while the designer may then have the maximum freedom to optimise the layout, the consequence is long development times and a susceptibility to errors.

At the other extreme there are design systems which treat ICs as if they were glorified printed wiring boards. The designer feeds in a logic diagram or similar structural description and some cpu time later an automatically placed and routed standard cell layout or gate array pops out. Such systems are sometimes called (for no obvious reason) silicon compilers. This approach, while it may reduce design time, sacrifices a considerable amount of silicon area.

Symbolic design takes a middle road between these two extremes. The intention is that the machine should do what the machine is good at (that is, handling the prodigious volume of detail) leaving the designer free to take the global decisions using his *unrivalled* intuition.

Symbolic design methods have been around for a number of years. In fact Larsen (1971) reported a quite advanced system in which a symbolic plot was used as the feedback mechanism in an automated layout approach.

Summarising the three main classes of symbolic methods that have been used for design at the cell level.

7.2.1 Coarse grid methods

In the coarse grid method, a cell is divided with a uniformly spaced grid in x and y. The spacing of the grid is such that most of the layout design rules are obeyed. Symbols are defined for all the possible components that may be placed on the grid and the layout method consists of 'tiling' the cell with these symbols to construct the required circuit. Gibson and Nance (1976) described probably the first system generally in use which employed this method. The approach, while increasing the layout engineer's productivity, obviously sacrifices significant silicon area and does not guarantee design rule correctness either from a layout or an electrical standpoint. However, coarse grid has remained popular, with at least two CAD manufacturers marketing systems.

7.2.2 Gate matrix
In 1980, Lopez and Law described a method called 'gate matrix' which is in essence a coarse grid approach. However, the placement of components is structured so that it is possible to place the grid lines much closer together than with the conventional coarse grid. In the gate matrix scheme polysilicon is constrained to run in one direction at a fixed pitch and width while diffusion runs orthogonally or in parallel in the 'non-poly' slots. Intersections of poly and diffusion form the transistors and metal is used (running in either direction) to complete the circuit (see Fig. 7.2). With this method packing densities close to that of hand packed layout have been demonstrated.

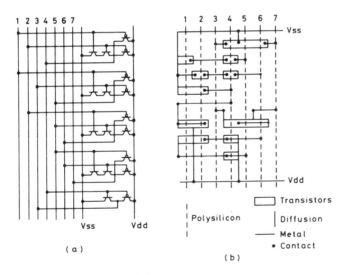

Fig. 7.2   *An example of a gate matrix layout*

In gate matrix, as described by Lopez and Law, the symbolic input method employs ASCII characters typed on a conventional text terminal and is rather cryptic to the unskilled.

7.2.3 Sticks
Symbolic design received a major fillip with the Caltech Silicon Structures Project (SSP) in the late 1970s and the establishment of the 'sticks' notation

originally developed by Williams (1977, 1978) of MIT. In 1980 'The Proposed Sticks Standard' was published (Trimberger, 1980) and Mead and Conway (1980) made extensive use of the notation. In the early work the sticks notation was mainly a convenient form for drawing circuit diagrams that conveyed some information about the physical characteristics of the circuits being designed. In parallel with this use, a number of systems were developed that allowed direct conversion of the sticks or 'sticks like' topology into compact mask geometry (Williams, 1978; Mosteller, 1981; Dunlop, 1980; Hsueh and Pederson, 1979; Weste, 1981). An example of a sticks layout (in a non-standard notation) is shown in Fig. 7.3.

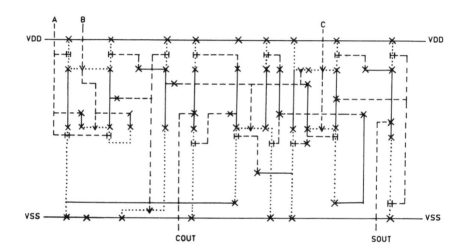

Fig. 7.3   *An example of a sticks layout*

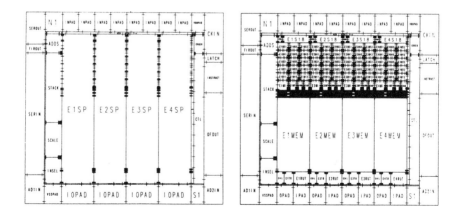

Fig. 7.4   *A floor plan at (a) an intermediate (b) a lower level in the hierarchy*

7.2.4 Floor plans

At a higher constructional level the use of floor plans as a symbolic notation was also developed at the SSP (Mead and Conway, 1980). Initially these were again used as graphical abstractions of the physical characteristics of the circuits being designed. Even today floor plans used for this purpose frequently have only a single hierarchical level. Only a few systems exist which use floor plans in a similar role to stick diagrams: for the creation of compact layout of large modules or complete chips (see, for example, Ackland and Weste, 1983; Revett and Ivey, 1983). In this extended use of floor plans it is essential that they are hierarchical so that the block topology of a large chip can be viewed without overwhelming the designer in detail (Fig. 7.4).

7.3 The objectives of symbolic design

What are the major objectives and motivations of a symbolic design system? The first objective of all CAD systems is to enable overall design times to be minimised. An important codicil to this objective (in an industrial context) is that design times should not be reduced at the expense of poor silicon area utilisation although some area increase is often acceptable (Fig. 7.5).

Secondly, in a world in which significant fabrication process evolution may take place during the design of a chip or system it is necessary to be able to

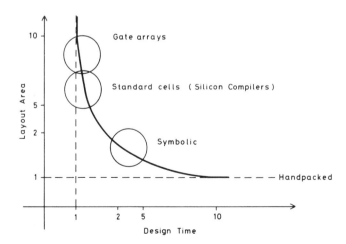

Fig. 7.5 *Silicon area utilisation for systems applications as a function of design approach*

follow any improvements in and recover gracefully from any deterioration in the design rules. We therefore need to be able to automatically update design data to accommodate changes in the design rules.

A third objective of a symbolic design system is to mimimise the amount of detailed knowledge required by designers and handle automatically (as far as possible) conformance with the design rules.

Fourthly, symbols may be used to abstract away from the detail of the implementation while retaining adequate information for appropriate decisions to be made at the level of interest.

How do symbolic design systems meet these goals and what are the criteria that must be satisfied? The first objective must be met by any CAD system if it is to be successful in the longer term. In addition to this, in many industrial environments the intention of CAD systems is to increase the productivity of expensive and scarce engineering staff and not, for example, to move jobs from the drawing office to the engineering lab. Much of the software for the current generation of CAE workstations comes into this category and, for example, a schematic capture terminal on every engineer's desk may, in fact, reduce productivity. This is especially true if schematic capture is used purely for documentation or solely as an input to a simulator since the same structure will need to be entered again when the chip is laid out. Symbolic design systems can overcome this problem by using the structural descriptions provided by the floor plans and the stick diagrams to generate the simulation data directly. The additional information provided by the topology inherent in the floor plans and sticks is, of course, used to generate the layout. This eliminates the need to enter a structural description of the chip twice with all that entails in terms of consistency checking and potential for errors.

Automatic updating of design rules is straightforward in a system in which the basic design data contains no dimensional information. The conversion programs that transform the topological descriptions to mask geometry have as one input a set of design rules and by modification of these tables a chip may be re-engineered over a wide range of silicon processes. There are of course limitations and the topological features of the processes must be compatible. For example, one could not suddenly introduce another layer of metal and reassemble the chip but one could port a design from a 3 micron to a 2 micron technology of the same type.

The third and fourth objectives have much in common with each other and broadly explain the use of symbolic cell methods and floor planning respectively. Symbolic systems of all types alleviate the need for detailed design rule knowledge although coarse grid systems handle conformance with the design rules in rather a crude way. On the other hand, a sticks system with compaction eliminates the need for design rule knowledge although it may in some measure replace this with a need to 'know your compactor'. Conformance with the design rules is clearly complete and the final layout is as close to minimum as the authors of the programs are able to make it!

7.4 Floor planning and high level design

The conventional system block diagram is the means by which most practising engineers describe the structural component of a high level design. Such block diagrams are usually hierarchical for a large system. Block diagrams are, of course, a symbolic description of the system but, for the purposes of custom VLSI design, important elements have been discarded in the abstraction process. The criterion mentioned above for a good symbol was that it should retain enough information to facilitate the decision making process at the current working level. In the case of VLSI design the important, indeed the only, explicit level of the design that *MUST* exist is the silicon level. In addition silicon area is limited and expensive (and will be for a long time to come!) so that a necessary parameter in even high level decisions must be an estimate of the silicon area that the element will occupy. Another important feature of the symbols must be how each element will 'jigsaw' into the final chip. As implied above the intention of symbolic systems is to manage the problems of chip design and allow the designer to make the decisions of a global nature. Included in these decisions is the overall wiring strategy of the chip together with the global and local signal routing.

High level design symbols must, therefore, represent a structural partition's size, shape and position on the final chip together with the more usual elements of a conventional block diagram notably interconnections. A hierarchical set of floor plans satisfies these criteria and can therefore be employed as a VLSI block diagram.

The form of a symbolic VLSI system at the higher levels of design would be such that the hierarchical floor plans replace the conventional block diagram and a floor planning tool replaces the schematic editor. For completeness each partition in the floor plan at any level in the hierarchy should have a behavioural specification preferably in a language which is simulatable. By this means a structural decomposition of the high level behaviour can be carried out in a generally 'top down' direction until a level is reached at which it is possible to implement the remaining partitions as interconnections of transistors.

7.5 Symbolic cell design

We reviewed the basic methods of symbolic cell design in section 7.2. In a modern symbolic system the sticks method with compaction is the one most likely to be used. The exact form of the symbols, however, depends upon the purpose of the designer at the time. If, for example, the designer is concerned mainly with the performance of the cell at a switch level and the consideration of the layout is of secondary importance then the conventional sticks diagram as described by Mead and Conway is the most appropriate. However, at a later stage in the design process, circuit performance and layout optimisation are uppermost and the 'logs' representation (Weste and Ackland, 1981) carries the additional information to aid the designer to make the necessary trade offs (Fig. 7.6).

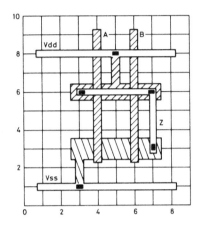

Fig. 7.6   *A 'log' plot (after Weste and Ackland, 1981)*

7.6 Algorithms for spacing

Once the cell toploogy has been generated in a sticks form it is necessary to 'expand' the data to take account of the design rules that prevail and so produce a mask description. This process is usually called 'compaction' for historical reasons. A more appropriate term might be spacing since the process usually consists of giving a non-metric description physical dimensions so that it occupies 'space' in 2 dimensions.

7.6.1 Early Compactors
Early work was carried out by Akers et al (1970) in which they used a compaction algorithm to remove waste area in a coarse grid layout. The algorithm worked on a matrix of elements which represented the tiles in the coarse grid topology. The method was to search for 'shear lines' that allowed space to be removed from the layout (see Fig. 7.7). Heuristics were used to ensure that the topology of the circuit was not altered in illegal ways.

The basis for the compaction of sticks symbols was described by Williams (1978). The essence of the spacing method is that the stick diagram is composed of relocatable elements wired together with stretchable interconnections. As long as the elements are free to move and the connections are infinitely stretchable a design rule correct layout is always possible.

Dunlop (1978) described a scheme, to arrive at a legal set of spacing, in which symbols and interconnections were manipulated by methods similar to those used in design rule checking programs. The compactor used Akers' algorithm to search for wasted space and achieved very good packing density but at the expense of long run times.

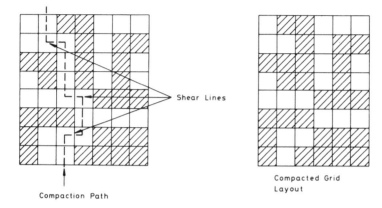

Shear Lines

Compaction Path

Compacted Grid
Layout

Fig. 7.7   *Use of shear lines in compaction*

7.6.2 Graph based method

Hseuh and Pederson (1979) proposed a graph algorithm in which the symbolic description is partitioned into vertically and horizontally connected features which are located on common centre lines. Graphs are then constructed, for each dimension, in which the nodes are these partitions and the edges are the constraints between them (Fig. 7.8). The critical path is then determined and used to derive the mask coordinates of each of the partitions. The process is repeated in orthogonal directions until convergence. This method was also used in SLIM (Dunlop, 1980) together with jog insertion and routing to alter the original topology in order to extract space.

7.6.3 Virtual grid

In 1981, Neil Weste of Bell Labs proposed an algorithm in which the symbolic circuit is partitioned with all the elements lying on a given grid line forming the node in a graph and the constraints between any element and all other elements forming the edges. An additional limitation is that the minimum grid spacing is zero so that grid lines may not cross even though there are no constraints between them. The spacing of the grid is simply determined by taking the worst case edge between any pair of grid lines. No form of processing that alters the topology is included.

The advantage of this method over conventional graph methods, especially those with automatic jog insertion etc., is that the results of changes made by the designer are predictable and interaction with the input symbolic data can be used to optimise the result. Further the designer's 'intent' is not destroyed by the virtual grid method – a feature which can avoid considerable frustration when using a CAD tool.

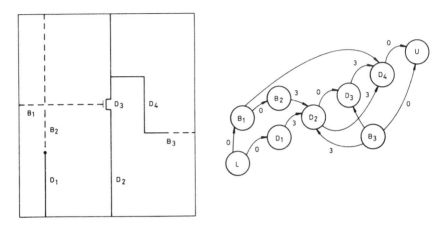

Fig. 7.8   *A symbolic layout and the 'x' constraint graph*

7.6.4 Recent developments
The methods described so far are all for one dimensional compaction. Recently there have been a number of improvements (for example, Boyer and Weste, 1983) published but these are of a fairly minor nature and do not affect the basic philosophy.

There has, however, been some attention given to the problem of simultaneous 2-D compaction. The problem was proved to be NP-complete by Sastry and Parker in 1982 but some heuristic methods have been published (Wolf, 1984; Kedem and Watanabe, 1983; Schlag et al, 1983). 2-D compaction should allow better methods of jog insertion (if that is desirable) and produce highly compact layout.

The final hurdle to producing layout as good or better than can be done manually is in the use of non-Manhattan geometries. To my knowledge no such compactors exist. A fruitful avenue of research should be along the lines of a 'constructive' design rule checking program. Such a program would be very expensive in computer time but with the aid of a hardware accelerator fast run times should be possible and the resulting layout very good.

7.7 Chip assembly

When the lowest level symbolic stick design and compaction of all the cells in a given module is complete, it is likely that they will not fully 'tile' the area assigned to them (Fig. 7.9) and that the pins connecting adjacent modules will not pitch match.

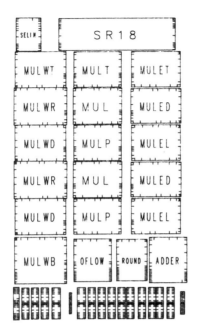

Fig. 7.9   *Compacted cells do not necessarily tile the full area and the pins do not pitch match*

I will refer to the process of 'assembling' a mask description from a set of floor plans and basic cells as *chip assembly* although this term may be used with other meanings which I will not pursue here.

There are three basic methods of chip assembly:

(i)   redesign the cells so that they do fully tile the space and pitch match,
(ii) route the interconnections and 'fill' the remaining area with 'white space',
(iii) stretch the blocks so that they pitch match and fill the space.

Method (i) is not appropriate to a symbolic system but is the method that may be used in a manual design approach. Methods (ii) and (iii) can and may be used. A number of workers have shown that method (iii) frequently produces the best results (Mudge, 1980; Weste and Ackland, 1981). A consequence of this method is, however, that a cell that is placed multiply in a design will in general produce multiple mask layout descriptions unless it is placed in exactly similar environments. This can lead to an explosive increase in the size of the mask database if care is not taken. In addition there may be certain components on a chip which have critical performance requirements such that stretching is undesirable (e.g. RAM cells). In these cases it may be necessary to adopt method (ii). Further, on any chip there are elements that have not been designed symbolically such as alignment marks, special clock generators with serpentine transistors etc. Obviously such elements cannot be stretched and must be routed into place.

A chip assembler must, therefore, provide both method (ii) and (iii) at the choice of the designer.

7.8 Algorithms for abutment

The algorithms for chip assembly have much in common with those for spacing discussed in section 7.6. In the case of chip assembly the fundamental symbolic entities are interconnection points between blocks and block edges. The constraints are provided by the edge to edge spacing required to satisfy the design rules and the need to match pins.

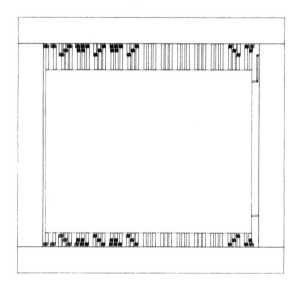

Fig. 7.10   *River routing around a rigid block*

The simplest algorithm that could be used is virtual grid. However, because the virtual grid maintains the grid order of the original symbolic design (the floor plans) it is impractical since it is difficult (not to say impossible) to predict the exact order of pins of, say, 2 blocks on opposite sides of a 7 mm wide chip. With virtual grid, therefore, inevitable errors in the pin order cause unacceptable stretching.

The critical path method described in 7.6 is much more successful since unconstrained pin pairs and edges are allowed to pass each other so that the ordering in the floor plan is not significant (Ivey and Revett, 1984). A problem arises, however, when it is necessary to include fixed geometry cells in a chip otherwise assembled by abutment. Such cells, in general, will either be too small or too large for the space allotted by the assembler. The solution in the former case causes few problems since the assembler can take account of the disparity by stretching surrounding cells. However, if the cell is too small (or the spacing of a pair of pins in the fixed cell is closer than in the stretchable cells) it is necessary to river route the component into place (Fig. 7.10). The assembler is then required to estimate the wiring space necessary to complete the routing and the classic problems of 'place and route' programs can arise. Fixed cells should therefore be used sparingly.

7.9 Conclusions

Chip design has significant qualitative and quantitative differences from older, more established printed wiring technologies and hence requires a different approach. Symbolic design can be used to advantage at all levels of the design hierarchy if the symbols are formulated to convey enough information to facilitate the choices being made.

The symbolic method can endow a design with a large measure of design rule independence and hence give a greater longevity and robustness to what otherwise might prove fatal changes in physical design parameters.

The main advantage of symbolic techniques is that they have the potential to maintain the high packing density of manual methods with few restrictions on the type of chip architectures employed but with much enhanced productivity.

References

ACKLAND, B., and WESTE, N. H. E.: 'An automatic assembly tool for virtual grid symbolic layout,' Proc VLSI 83, Trondheim, Norway, Eds. Anceau and Aas, North Holland Publishing (August, 1983)

AKERS, S. R., GEYER, J. M. and ROBERTS, D. L.: 'IC mask layout with a single conduction layer', Proc 7th Annual Design Automation Workshop, San Francisco (1970)

BOYER, D. G. and WESTE, N. H. E.: 'Virtual grid compaction using the most recent layers algorithm,' International conference on computer aided design, Santa Clara, CA (September, 1983)

DUNLOP, A. E. 'SLIP: Symbolic layout of integrated circuits with compaction' Computer aided design, **10** No. 6 (November 1978)

DUNLOP, A. E. 'SLIM – The translation of symbolic layouts to mask data', Proc. 17th Design Automation Conference (June, 1980)

GIBSON, D. and NANCE, S. 'SLIC – symbolic layout of integrated circuits', Proc. 13th Design Automation Conference (June, 1976)

HSEUH, M. Y. and PEDERSON, D. O. 'Symbolic layout and compaction of integrated circuits', Proc. International Symposium on Circuits and Systems (July 1979)

IVEY, P. A. and REVETT, M. C. 'Custom VLSI design using the ASTRA CAD system', Proc. of the European Electronic Design Automation Conference (March 1984)

KEDEM, G. and WATANABE, H. 'Graph optimization techniques for IC layout and compaction', Proc. 20th Design Automation Conference (June, 1983)

LARSEN, R. P. 'Computer aided preliminary layout design of customized MOS arrays', IEEE Transactions on Computers, vol. C-20 No. 5 (May, 1971)

LOPEZ, A. D. and LAW, H.-F. S. 'A dense gate matrix layout method for MOS VLSI' IEEE Journal of Solid State Circuits vol. ED-22 No. 8 (August, 1980)

MEAD, C. and CONWAY, L. 'Introduction to VLSI systems', Addison Wesley Publishing (1980)

MOSTELLER, R. C. 'REST a leaf cell design system', Proc. VLSI 81 Ed. J. P. Gray, Academic Press (August, 1981)

MUDGE, J. C., PETERS, C. and TAROLLI, G. M. 'A VLSI chip assembler', Design Methodologies for VLSI, NATO Advanced Summer School, Louvain-la-Neuve, Belgium (July 1980)

REVETT, M. C. and IVEY, P. A. 'ASTRA – A CAD system to support a structured approach to IC design'. Proc. VLSI 83 Eds. Anceau and Aas, North Holland Publishing (August, 1983)

SASTRY, S. and PARKER, A. 'The complexity of two dimensional compaction of VLSI layouts', International Conference on CAD, New York, NY (1982)

SCHLAG, M., LIAO, Y. Z. and WONG, C. K. 'An algorithm for the optimal compaction of VLSI layout', International Conference on CAD, Santa Clara, CA (September, 1983)

TRIMBERGER, S. 'The proposed sticks standard', Caltech Technical Report, TR 3880 (October, 1980)

WESTE, N. H. E. 'MULGA – An interactive symbolic layout system for the Design of integrated circuits', Bell System Technical Journal, vol. 60 No. 6 (July–August, 1981)

WESTE, N. H. E. 'MULGA – An interactive symbolic layout system for the design of integrated circuits', Bell System Technical Journal, vol. 60 No. 6 (July–August, 1981)

WILLIAMS, J. D. 'Sticks – A new approach to LSI design', Masters Thesis, Massachusetts Institute of Technology (June, 1977)

WILLIAMS, J. D. 'Sticks – A graphical compiler for high level LSI design', Proc. NCC, vol. 47 (June, 1978)

WOLF, W. 'Two dimensional compaction strategies', Ph.D. Dissertation, Stanford University (1984)

Acknowledgement

The author would like to thank the Director of Research, British Telecom for permission to publish this work.

Chapter 8

IC layout verification

R. A. Cottrell

8.1 Introduction

The final product of a CAD system for IC design is mask layout data. The subsequent steps required in the production of ICs belong to fabrication rather than design. It is therefore essential, if the circuits are to function correctly, that these data should be error free. The cost of a single fabrication run is so high that it is worth expending considerable effort and consuming large quantities of computer time to ensure that this is so.

If the mask layout data is to be error free, then it follows that all other data associated with the design, which has been used to derive the mask data, must also be error free. If the logic circuit does not correctly implement the specification, or the formal specification itself does not represent the designer's intentions, then the probability of obtaining correctly functioning ICs is infinitessimal, thus the provision of simulators and other tools operating at various levels of abstraction is essential. IC layout verification tools are intended to ensure that the mask data is a correct implementation of the circuit which must previously have been verified.

8.2 Automated design techniques

Some of the proponents of automated design techniques such as silicon compilation, argue that such techniques obviate the need for traditional mask data analysis and verification tools. Designs produced in this manner will, it is argued, be 'correct by construction' and need not be checked. However, it is a brave person who claims that their software is infallible, and checking can therefore provide a safety net. In addition, even in automatic systems it is desirable to feed back information on wire delays, etc., to permit more accurate simulation. In any case, it is not the author's opinion that automated techniques will wholly replace hand-crafted layout within the forseeable future.

Additionally, all automatic techniques require some previous hand layout. For example, a standard cell system requires the basic cells to have been laid out, and these are essentially full-custom hand-crafted designs. Whereas these will have been verified and characterised by fabrication, layout verification tools would still be valuable at the design stage.

8.3 Analysis tools for semi-custom design

Semi-custom ICs fall largely into two groups: gate arrays and standard cell based systems. In both cases, the design task consists of placing the logic functions within the chip and routing between the various blocks, both of which may be performed manually, fully automatically or with some interaction between man and machine. In both cases the basic building blocks may be assumed to function correctly as this will have been verified previously by actual fabrication, and the task of verification thus consists of verifying the interconnect. There are basically two layout verification tools required for semi-custom design: circuit verification and wire delay extraction.

8.3.1 Circuit verification
The purpose of a circuit verifier is simply to ensure that the layout is a correct implementation of the circuit diagram. It is not strictly necessary where the layout has been generated fully automatically, although it can still provide additional verification. It is absolutely essential when any kind of manual intervention in the routing task has taken place.

Circuit verification tools vary considerably in complexity. Some require the designer to give a corresponding name to all components in the circuit and layout databases, thus limiting the task to one of checking interconnections. Others require only the inputs and outputs of the circuit to be named, the mapping of components in the two databases being performed by the program. Naturally, these latter tools represent a significantly greater software investment and use up more computer resources.

8.3.2 Wire delay extraction
Whereas a layout may be a functional implementation of a particular circuit, it is also important to know the speed of operation. Wire lengths in semi-custom designs can be quite long, having significant resistance and capacitance and thus causing significant delays. It is important that simulations should be

performed with delays specified as accurately as possible, thus requiring the extraction of wire delays from the layout. Some such extractors simply calculate the capacitance of the tracks from their total area. This is adequate provided that the impedance of the driving circuits is significantly higher than the wire resistances. However, with increasingly large and complex circuits, wire resistance is becoming a more important factor which cannot be ignored. Wire resistance is more difficult to extract than capacitance, as it involves the estimation of the length to width ratio of wires rather than only their areas.

8.4 Analysis tools for full-custom designs

It is in the case of full custom, where a human is responsible for design right down to the transistor level, manipulating shapes on various layers, that layout verification tools are absolutely essential.

8.4.1 Dimension rule checking

All IC fabrication processes have associated with them a set of dimensional rules for the various masks. Such rules may be simple, involving only one layer, or more complex involving shapes on two, three or even four masks. Simple rules specify, for example, the minimum separation between two shapes on the same layer. An example of a more complex rule would be one which specified a minimum separation between a contact to polysilicon and active area. See Fig. 8.1 for diagrams of these examples.

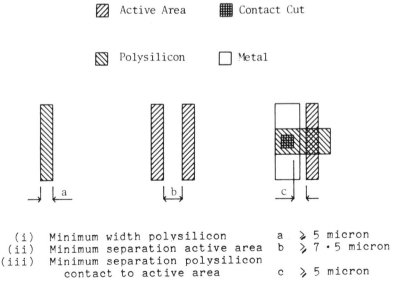

```
(i)    Minimum width polysilicon          a  ≥ 5 micron
(ii)   Minimum separation active area     b  ≥ 7·5 micron
(iii)  Minimum separation polysilicon
         contact to active area           c  ≥ 5 micron
```

Fig. 8.1 *Examples of design rules*

Dimension rule checkers were the earliest layout verification tools available. They operate on the layout database alone, without reference to the circuit. A typical dimension rule checker is technology independent, with the rules being written in a special language. This language will permit topological, set and quantitative operations to be performed on the basic shapes which comprise the layout. See Figs 8.2, 8.3 and 8.4 for examples of such operations. Once the rules have been composed, they are then compiled into an internal format before being applied to the layout. Typically, output is in two forms: a textual list of errors and their locations, and graphical markers on the layout suitable for display on a graphics terminal or pen plotter. Examples of the input and output for the rule of Fig. 8.1 (iii) are given in Fig. 8.5.

Dimension rule checkers simply verify that the layout is legal. No attempt is made to assess whether the layout is a valid representation of any particular circuit, nor even whether the devices are connected together sensibly to form legal logic gate circuits.

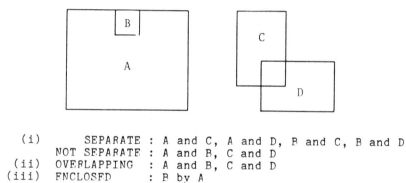

```
(i)       SEPARATE     : A and C, A and D, B and C, B and D
          NOT SEPARATE : A and B, C and D
(ii)      OVERLAPPING  : A and B, C and D
(iii)     ENCLOSED     : B by A
```

Fig. 8.2   *Examples of topological operations*

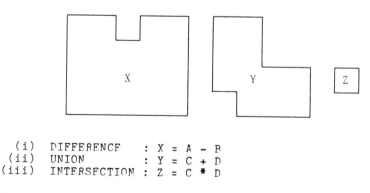

```
(i)    DIFFERENCE    : X = A - B
(ii)   UNION         : Y = C + D
(iii)  INTERSECTION  : Z = C * D
```

Fig. 8.3   *Examples of set operations*

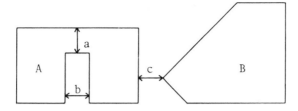

```
(i)     WIDTH OF A          = a (narrowest point)
(ii)    INTERLIMB OF A      = b (narrowest point)
(iii)   SPACING OF A and B  = c (nearest point)
```

Fig. 8.4 *Examples of quantitative operations*

8.4.2 Circuit extraction

The aim of circuit extraction is to convert mask layout data into electrical circuit data. The circuit obtained may be used directly as input to a simulator such as SPICE, or may be verified against an already designed circuit. The program must first recognise the devices: transistors, contacts, etc., from which the design is constructed (Dobes and Byrd [1]). This can be achieved largely using the same basic shape operations as for dimension rule checking. Once the devices have been recognised, their interconnection may be deduced and the circuit extracted. Some circuit extractors go further than generating the circuit in terms of transistors, and attempt to reconstruct the circuit in terms of logic gates. For MOS circuits, this consists largely of searching for groups of transistors which form a current path from VDD to ground (see Fig. 8.6). Simple NMOS NAND and NOR gates are easy to recognise; it may be necessary to resort to graph theory for complex CMOS gates.

8.4.3 Electrical rule checking

Electrical rule checkers operate on circuits extracted from layout and verify conformity to a number of electrical design rules. They vary in complexity from simple connectivity checkers to quite complex and thorough verification tools. The kind of checks which can be made are largely designed to ensure that the circuit is sensible. For example, in an NMOS process, it may be checked that all pull-up devices are depletion mode and that appropriate aspect ratios are used for pull-up and pull-down devices.

8.4.4 Netlist to layout verification

The purpose of such a program is much the same as the circuit verifiers for semi-custom design: to ensure that the layout is a valid representation of a circuit already designed and verified (Watanabe et al [2]). It will take the circuit extracted from the layout and compare it with the correct circuit. The user is required to provide a few 'seeding' names to give the program the correct starting point, typically the external connections. The algorithms of circuit comparison are such that execution times are usually reasonably short if the circuits are identical. However, it can take a long time to prove that two circuits are different, because every possibility must be investigated, however improbable.

```
RULE ONE
AA IS SHAPE MASK 2
POL IS SHAPE MASK 6
CUT IS SHAPE MASK 8
PC = CUT*POL
FAIL 'SEPARATION POLY CONTACT FROM ACTIVE AREA' &
    IF SPACING (PC, AA)<5 OR NOT SEPARATE (PC,AA)
END
```

(i) Rule described in a design rule language

```
!! COMMENCING APPLICATION OF RULE: ONE
PROCESSING Main Definition
SEPARATION POLY CONTACT FROM ACTIVE AREA
  RULE: ONE   VIOLATION 1
    VIOLATION OCCURS AROUND: X =   2·500 =   17·500
    RECTANGLE AT:
    MINX = 2·5   MINX = 17·5
    MAXX = 7·5   MAXY = 22·5
    RECTANGLE AT:
    MINX = −10   MINY = 10
    MAXX =   20   MAXY = 15
!! APPLICATION OF RULE: ONE   COMPLETE
NUMBER VIOLATIONS FOR THIS RULE   1
NUMBER OF SHAPES LOADED FOR RULE   4
```

(ii) Textual output from design rule checker

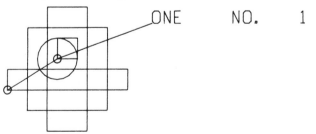

(iii) Graphical output from design rule checker

Fig. 8.5 *Example of using a design rule checker for the rule of Fig. 8.1 (iii)*
(Examples used by permission of Prime Computer CAD/CAM Ltd.)

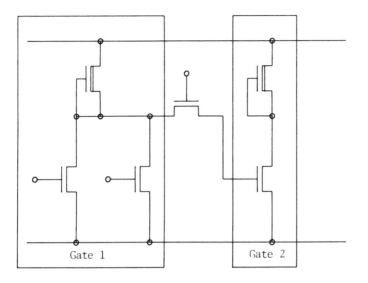

Fig. 8.6   *Grouping transistors into gates*

8.4.5 Parameter extraction

Parameter extraction will often be carried out at the same time as circuit extraction. Certain parameters, such as the width to length ratios of transistors, are required for electrical rule checking.

The parameters referred to are largely parasitic values such as the resistance and capacitance of interconnecting media. This process is therefore similar to the wire delay extraction for semi-custom circuits (qv Section 8.3.2) and has the same aim: the obtaining of more accurate simulation results. For this reason, it is important that the parameter extractor should be able to 'back-annotate' its results into the database containing the corresponding circuit description.

The extraction of capacitance parameters is fairly trivial, as capacitance is simply a function of the areas and perimeters of shapes on various layers. Indeed, where active area is concerned, SPICE takes the area and perimeters of the diffusion as the relevant parameters and calculates the capacitance values itself as these are voltage dependent. It is the extraction of resistance which is much more difficult (Horowitz and Dutton [4]), but fortunately accurate values of resistance (except in terms of channel lengths) are usually not required as track resistance is a second-order effect. Exact resistance extraction requires the solution of Laplace's equation by numerical means, which is a slow process. Where a polygon on a particular layer has more than two points of contact, it is necessary to solve Laplace's equation for various different voltages on the contacts as the equipotential lines will not always be the same shape. See Fig. 8.7 for an example of this. In general, some heuristic algorithm must be selected for the extraction of resistance information. This will generally consist of breaking complex polygons down into rectangles whose resistance can more easily be estimated. The problems is made much worse if non-orthogonal shapes are involved.

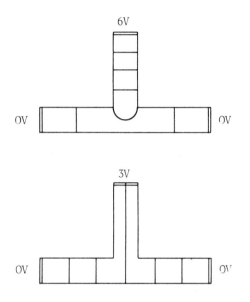

Fig. 8.7   *Approximate equipotential patterns for different voltages applied to the same shape*

8.5 Exploiting hierarchy

All VLSI IC designs must exploit hierarchy (Newell and Fitzpatrick [3]; there is no other way to manage the complexity required. In circuit terms, this means that gates are defined in terms of transistors, subsystems in terms of gates and systems in terms of subsystems. In layout terms, cells are defined in terms of shapes on various mask layers, these cells can then be incorporated with other cells and primitive shapes to form larger cells and eventually complete ICs. Clearly, if layout verification tools are to be useful for complex ICs, then they must be able to exploit this hierarchy if computer execution times are not to be excessive. The author has performed dimensional rule checks without hierarchy on relatively simple designs, incorporating a few thousand transistors, which have taken in excess of 10 hours of CPU time on one of the fastest minicomputers available today. However, exploiting hierarchy in layout verification is not simple.

8.5.1 Overlapping cells and boundary problems

Where cells are allowed to overlap with other cells or primitive shapes, problems are caused for all layout verification tools. Problems can also arise, particularly in the case of dimension rule checking, even where cells only abut or are in close proximity. A cell may pass the design rule checker internally and yet failures may be produced when it is laid down in proximity to another cell. Equally, it is possible for a cell to show a design rule violation which is removed by being overlapped with or abutted to another cell. The problem is similar for circuit extraction; overlapping or abutting cells may alter the circuit. In fact, it is only by cells abutting or overlapping with other cells or primitive shapes that the overall circuit can ever be extracted.

Overlapping cells are a real problem and cannot easily be accommodated by layout verification tools. One solution is simply to fully instantiate those parts of a design where cells overlap, but this may result in the obliteration of most of the hierarchy in a number of designs. One alternative is to re-map the cell boundaries to produce new, non-lapping cells. This is illustrated in Fig. 8.8. However, this approach introduces more problems of its own, not the least being that active devices may be cut down the middle. It will also ensure that the layout and circuit hierarchies do not correspond. A more promising approach seems to be to establish whether the overlap region causes any significant alteration to either cell. If it does not, it can be ignored; otherwise that part of the design must be instantiated.

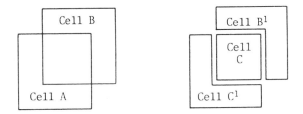

Fig. 8.8   *Re-map cells to remove overlap*

Once a situation is achieved where cells are allowed only to abut and not overlap, the problems may largely be reduced to intra-cell and inter-cell checks, although it would be naive to suggest that this simple step will obviate all problems concerned with exploiting hierarchy. However, it does represent a significant step in the right direction.

8.5.2 Correspondence between circuit and layout hierarchies

Quite simply, if the hierarchical structures of the circuit and layout databases do not correspond exactly, then the tasks of netlist to circuit verification and the back-annotation of extracted parameters will be impossible. It is therefore up to the designer to ensure that this is the case. He may use intermediate levels of hierarchy in either database, but the points of correspondence between the two databases must be clearly defined if hierarchy is to be exploited in the tasks of netlist to circuit verification and parameter extraction.

Clearly the re-mapping or instantiation of overlapping cells as proposed to deal with overlapping cells will interfere with this hierarchical correspondence. The designer must ensure that he does not use overlapping cells at a level within the hierarchy where he desires the two databases to correspond.

8.6 Databases for IC design

The form of database used in IC design can have a significant impact on the ease or difficulty in performing layout verification tasks. The presence or absence of certain pieces of information can be vital as can the form in which the data are stored. The incorporation of all information, circuit and layout, into one database can improve the overall integrity of the system.

8.6.1 The case for a unified database
If all the data associated with an IC design are held in one unified database, then that database can keep track of which parts of the circuit have been verified. When a modification is made to any part of the circuit or layout, it can be noted in the database and further checks required. When a circuit is finally to be fabricated, the database can confirm that all parts of the circuit have been verified to the satisfaction of the designer.

In addition, the unified database concept can help the designer to ensure the circuit and layout hierarchies are made to correspond. In this way, large amounts of effort will not be spent in designing layout which cannot be verified against the circuit it was intended to implement. Indeed, it may be possible to store layout and circuit information in one hierarchy, with each object in the hierarchy referring to the layout and circuit associated with it.

8.6.2 Information required in the layout database
For the tasks of netlist to layout verification and parameter extraction, it is essential that signal names can be assigned to shapes within the layout to correspond with node names in the circuit database. In particular, the external description of a cell (i.e. how it looks from the outside rather than its internal workings) should include the cell boundary and the location, mask layer and signal name of all its external connections. These data are not available in a number of older, purely layout databases.

8.6.3 The advantages of a symbolic layout database
In a symbolic database, the primitive elements are transistors, contacts, wires etc., rather than basic shapes on various mask layers. This is clearly an

advantage in circuit extraction, as the process of device recognition is redundant. Indeed, as a symbolic database essentially combines layout and circuit information, it is much easier to achieve the goal of a unified database and a single hierarchy.

The process of dimension rule checking is also simplified because the symbols are known to be internally correct. The main checking which is still required is to ensure the designer does not place symbols too close together. Within an individual cell, this is simple enough to be performed in real-time as the designer uses a graphics editor, with only inter-cell checks required as a batch-mode operation. An alternative approach is for the designer to place symbols on a virtual grid; a batch mode compactor then attempts to place the symbols as close together as possible to minimise silicon area. In this way, no design rules are violated and the layout is correct by construction.

8.7 Conclusions

The primary function of IC layout verification tools is to verify the correctness of the mask layout data with respect to the circuit data. Some tools simply establish whether the layout is legal, whereas others check the layout against the circuit and can insert into the circuit database parametric information extracted from the layout.

It is highly desirable to exploit hierarchy in layout verification, although there are serious problems which can make this difficult. Techniques are available to circumvent some of these, but much responsibility lies with the designer. The use of a unified database for both layout and circuit information is useful in enabling the exploitation of hierarchy. The inclusion of circuit-related data such as node names in the layout database can aid checking programs. The storage of layout data in symbolic form rather than basic shapes is also worth consideration, as it provides significant help with all tasks associated with layout verification.

References

1 DOBES, I. and BYRD, R.: 'The automatic recognition of silicon gate transistor geometries – an LSI design aid program', 1976, 13th Design Automation Conference Proceedings, 327–35.
2 WATANABE, T. *et al.*: 'A new automatic logic interconnection verification system for VLSI design', IEEE Trans. Computer Aided Design of Integrated Circuits and Systems, 1983, CAD-2, 70–81.
3 NEWELL, M. E. and FITZPATRICK, D. T.: 'Exploitation of hierarchy in analysis of integrated circuit artwork', IEEE Trans. Computer Aided Design of Integrated Circuits and Systems, 1982, 192–200.
4 HOROWITZ, M. and DUTTON, W.: 'Resistance extraction from mask layout data', IEEE Trans. Computer Aided Design of Integrated Circuits and Systems, 1983, CAD-2, 145–150.

Chapter 9

High level languages in design

M. R. McLauchlan

9.1 Introduction

The arrival of Large Scale Integration forced circuit designers to review how designs were represented. Up to this time, designs were usually hand drawn, often on large pieces of mylar, using coloured pencils. Design changes resulted in the drawings being altered or re-drawn. Reviewing a design for errors or anomalies would involve a group of designers looking at the drawings in detail and would result, occasionally, in some errors being overlooked.

That this was possible was due to the simplicity of the circuits being designed. The drawings were never too complicated, because of the limitations of the technology in which they were to be built. Not until the arrival of the Integrated Circuit in the mid-1960's was there any real pressure to alter this approach to design 'capture'. Even then, with the severe limitations of Small and Medium Scale Integration, no real changes were seriously proposed.

However, now that Large and Very Large Scale Integrated Circuits can and are being designed, the inherent problems of graphical design techniques have become apparent. Attempts have been made (some very successfully) to automate the entire process of schematic capture. But it is still a highly complicated activity, requiring expensive special-purpose computers which provide semi-automatic layout, check for design rule violations, permit small changes to the artwork or allow extensive (and expensive) simulations of the entire circuit to be carried out. This is not to say that there is no place for graphics in the design process – far from it. Rather, the emphasis is now moving towards other, complementary notations, which if used in parallel with traditional diagrammatic representations of the circuit, will allow the designer to understand the design much better, as well as providing a greater degree of confidence in its overall correctness.

With the expansion in the late 1960's and early 1970's of the Computing Science community, and in particular, the proliferation of new and increasingly sophisticated Programming Languages, it was hardly surprising that the idea of using a 'language' to describe a circuit was born. In fact, it was already true that for SSI and MSI designs, the finished circuit diagram, which detailed the layout of the basic shapes necessary to construct the chip, had to be translated into a suitable format in order to drive the pattern generation machines that produced the masks. These masks were ultimately used to draw the geometric patterns onto the surface of the silicon wafer. At first, each different machine had its own 'instruction set', but this had the obvious handicap that a design translated into 'format A', could not be used on a machine that required 'format B' without either extensive modification, or indeed complete rewriting.

This simple problem triggered the design and implementation of a number of standard, machine readable notations, with which to represent the geometric forms of interest to ic-designers. CIF, as described in Mead and Conway (1), and GAELIC, as described in Mavor *et al.* (2), have both been used extensively in this role. Even today, these very simple languages are used as an 'intermediate format', thus allowing a wide variety of quite different processes to communicate. For example, plotters, video displays, pattern generators, E-beam machines, and design rule checking tools can all use such a standard notation (suitably translated) to carry out their appropriate function. These languages are very much the 'assembler languages' of the Integrated Circuit design world.

To give an example of what they look like, here is a short extract of a GAELIC program to describe a shift register cell, as found in (1).

```
NEWGROUP SRCELL;
TRACK (1) 2, S, 0, 6: 10;
TRACK (1) 2, S, 13, 0: 25;
TRACK (1) 2, S, 16, 6: 5;
RECTANGLE (1) 2, 11: 6, 7;
POLYGON (2) S, 3, 0: 4, 3, 1, 5, 3, 3, 5, −2, 4, 4, −11, −3, −1, 2, −1,
               2, −1, 5, 1, 4, −4, −4, 1, −5, 1, −2, −1, −9, 1, −3;
TRACK (3) 4, S, 0, 2: 21;
TRACK (3) 4, S, 0, 21: 21;
RECTANGLE (3) 3, 8: 4, 6;
RECTANGLE (3) 16, 7: 4, 6;
RECTANGLE (4) 2, 9: 6, 11;
RECTANGLE (5) 4, 1: 2, 2;
RECTANGLE (5) 4, 9: 2, 4;
RECTANGLE (5) 4, 20: 2, 2;
RECTANGLE (5) 17, 8: 2, 4;
ENDGROUP;
GROUP SRCELL, 0, 0, XX, X, 3, 20;
```

The Gaelic language allows designers to arrange basic geometric shapes in suitable patterns, on particular layers, in order to form the required circuit. What is more, it allows shapes to be grouped together to form more complicated structures, which when given a name, may be instanced and orientated in the most suitable way. The above example demonstrates quite adequately that programming in such a simple notation can be time consuming and error prone. It is far from obvious what is being done, and the only way progress can be made is for the program to be regularly compiled and the results displayed on a graphics terminal. It was quickly realised that if the promise of VLSI was to be fully exploited then languages such as CIF and GAELIC were not going to be adequate.

At about the same time, the interest of a large number of academic institutions both in the United Kingdom and abroad were aroused. A lot of researchers in Computing Science Departments in particular started investigating the possibilities of developing languages dedicated solely to the task of designing integrated circuits. It gradually became apparent that the following issues needed to be investigated, and where necessary, solved.

– if VLSI was to be exploited successfully, the complexities of the design process had to be solved,
– existing languages were inadequate,
– programs representing designs would be easier to modify than pictures,
– properly defined languages could be checked automatically for correctness,
– designs consisted of behavioural and structural information, as well as physical,
– previous research in the area of Programming Languages could be applied in this area,
– language descriptions of circuits could be quite concise, and therefore easily communicable between individuals, design teams and organisations.

The remainder of this chapter will look at how the above ideas developed, and will ultimately provide a glimpse of what the future may hold. Chapter 11 will investigate this subject further, by introducing and explaining the concept of Silicon Compilation.

9.2 First developments – Bristle Blocks

One of the first projects to explore the use of a high level language for designing Silicon, was the Bristle Blocks system, as proposed by Johanssen (3), which was developed at the California Institute of Technology.

The system was intended to provide a design environment in which LSI circuits could be designed without the need to know too much about the underlying theory. The hope was that an entire mask set could be produced from a single page, high level description of the required circuit.

This was to be achieved by providing the user with a primitive component called a cell, which could contain geometric primitives and references to other cells. A basic rule in Bristle Blocks was that a design would be expressed in a hierarchical manner, thus introducing a form of locality which could be exploited by design rule checkers and electrical simulators. Hierarchical design had been found to be crucial to good design. In a similar way, a structured design approach was encouraged and was based on the methods put forward in (1).

A variety of representations needed to be captured in a Bristle Blocks design, and these included the physical (e.g. layout), sticks, logic and textual forms. Unfortunately, these were captured separately, thus allowing inconsistencies to develop as modifications were made to the design.

A facility was provided to allow previously defined cells to be included within a new design. They were stored in libraries. This permitted the sharing of designs and was particularly worthwhile as it encouraged designers to use designs that had been previously tested and were known to be correct.

New low level cells could be defined using a suitable cell design language, for example CIF. It was felt that a designer would be better at designing these low level cells, as in general, they were not very complicated and took only a little time to implement. In addition, designer ingenuity might well succeed in reducing the size of a low level cell to a minimum, something an automatic generation system would have great difficulty in doing. An important difference between Bristle Blocks cells, and cells as found in languages such as CIF were that the Bristle Block cells were dynamic. That is, they could be used to do a variety of different things. For example, compute their power requirement, provide a textual description or simulate themselves. They could also draw themselves, deforming their shape if necessary. In contrast, CIF type cells were only ever able to do the same thing when requested in a design. That is, produce the layout of the cell. They were therefore static objects.

In the implementing of a chip, Bristle Blocks provided a fixed architecture. This required the definition of three different formats. These governed how cells would be arranged on the chip (otherwise known as the structural format), how they would interact with each other (otherwise known as the logical format), and how they would pass information between each other (otherwise known as the temporal format).

A Bristle Blocks compiler was implemented which, though quite simple and which only supported a sub-set of the envisaged tools, was used to generate a number of chips. The major limitation of the project was the machine on which it was initially implemented.

One of the significant aspects about the Bristle Blocks approach was that it introduced a new phrase into LSI (and VLSI) design. It was the term Silicon Compilation. The concepts suggested were very similar to those behind Software Compilation. They were, that from an initial high level description of what was required, a program could take that description and transform it into a form suitable for it to be 'realised' in an appropriate, implementable technology. For example, machine code. In a similar manner, a Silicon Compiler, from a high level description of the circuit required, would produce automatically a variety of information that could be used by the design system and designer to improve the design. Finally, it would produce the 'assembler/ machine code' needed to generate the masks for ultimate fabrication, with no interference from the user.

9.3 The influence of programming languages

9.3.1 Embedding physical information

A major disadvantage with the Bristle Blocks approach to LSI design, was that effort had to be expended in designing and implementing the language and associated design system. Many designers felt that another, alternative approach to the problem, would be to use existing High Level Programming Languages. By embedding a suitable interface into the chosen language, which essentially involved defining a number of different parameterised procedure calls, all the advantages of a high level language like conditional and repetitive statements, procedures, parameters, and the ability to define variables would be available to a designer used to using CIF or GAELIC. What was even more important was the ability for the 'design language' so constructed, to develop and grow in a manageable way, as more facilities were added to the interface.

For example, in the beginning, such a design language would contain procedures that upon being called, would generate the appropriate layout. The parameters would specify where the object was to be placed in two-dimensional space, plus other important facts about the object, like orientation, etc. Such objects might be basic logic components like AND or OR gates, input and output pads or similar. After a time, even more sophisticated devices might be defined, such a Programmable Logic Arrays (PLA's), or memories. The parameters would still specify their origins, but in addition more abstract information could be provided, such as the number of distinct inputs and outputs, the projected overall size of the device (for example, Number of Rows X Number of Columns), and the information needed to program such devices as PLA's and PROM's. After more development, even higher interfaces might be defined, providing the capability to route and place previously defined components. All of these facilities would be available from within the same base language thus avoiding the need to retrain the designer.

The first example of such a language was LAP, described by Locanthis (4), developed at the California Institute of Technology, and based on the programming language Simula. Other variations on the same basic idea have been ILAP, described in (5), which was developed at the University of Edinburgh and used IMP as the base language, and finally PLAP, documented by Fletcher and Mole (6) and developed at the University of Newcastle upon Tyne. It was based on the language Pascal.

To show what such languages were like, here are some extracts from a PLAP description of a Shift Register. A Pascal programmer should have no difficulties following the text, though one non-standard feature of the language is used to permit suitable libraries to be inherited when needed.

PLAP has been developed on a VAX-VMS system and the Pascal used in this example thus corresponds to the language

```
[INHERIT('$plap:.inc')]
PROGRAM shiftreg;
VAR i : integer;
BEGIN
    initialise('shiftreg');
    define–pads;

    define ('Srcell');
        rect(METAL,0,0,42,8);
        rect(METAL,0,38,42,8);
        .
        .
        .
        rect(CUTS,8,4,4,4);
        rect(CUTS,34,16,4,8);
    enddef;
        .
        .
    define('Shiftslice');
        instance('Longtrack','',maxint,0,0,xx);
        instance('Srcell','',maxint,0,53,xx);
        instance('Srcell','',maxint,42,106,y);
        instance('Srcell','',maxint,0,159,xx);
    enddef;
        .
        .
    define('Row');
        instance('Leftedge','',maxint,0,0,xx);
        instance('Leftextra','',maxint,28,0,xx);
        FOR i := 1 TO 10 DO
            instance('Shiftslice','',i,36+(i-1)*42,0,xx);
        instance('Rightedge','',maxint,36+(10*42),0,xx);
        instance('Padout','',maxint,36+(10*42)+290,0,rrr);
    enddef;
        .
        .
        .
```

```
define('Chip');
.

.
enddef;

finish('Chip')
END.
```

available on this machine.

This short example highlights the main shortcoming of the notation, which is its verbosity. The designer is preoccupied with placing the separate components of the design and it is also difficult to determine which components are connected and how. Routing can be performed either automatically using the appropriate procedure, or manually by the designer, thus allowing errors and inconsistencies to be introduced into the design.

9.3.2 Structural and behavioural issues

In an effort to tackle these problems, and in trying to solve some of the more fundamental problems inherent with VLSI design, the approach of using a high level language has been extended so that instead of just recording the physical design (that is, the placement of the constituent geometric forms that go to make up components), the program description captures the structural (that is, the decomposition of a design into simpler components and their interconnection) and functional (that is, the projected behaviour of each component) forms of the design as well. For a successful design, all three need to be thought out. Ultimately, the physical design is used to generate the mask layouts, but before that stage is reached it is useful to capture formally what the design is supposed to do and how by breaking it down into simpler parts it can be achieved. In software terms, the structural design is captured by using different procedures, functions, parameters and global variables and imposing an order on their use. The functional design is often just an idea in the mind of the designer. Rarely, when writing a program, is a programmer required to initially generate formal specifications. This state of affairs is, thankfully, being remedied, but it is interesting to note that even in these early days of designing VLSI Circuits, the need to record initial specifications has been identified as being of paramount importance.

Looking into the future, it can be envisaged that a designer will only need to give a formal specification of the circuit required, which would leave the design system to sort out the physical implementation. This obviously involves structuring the design and generating the layout. However in the short term, the whole process has to be the responsibility of the designer. Therefore, the design language must be able to capture all three forms of the design, and the overall design system needs to be able to manipulate the design information in such a way that the designer gets maximum benefit. For example, the functional information will no doubt be specified for all components of the system, though only one functional specification was provided initially (for the component being designed). Having recorded all this information, it could then be used to improve simulations of the component, by allowing the system to use, where appropriate, global specifications, instead of individual sub-component specifications, connected together as defined by the structure. As a consequence, the efficiency of the simulations could be improved, through better use of the machine and the improved interface provided for the designer.

In trying to implement the above ideas in a standard programming language, a similar approach as before is taken, of allowing the notation to develop through the addition of new features, so as to avoid the need for continual retraining of the designer. In addition, it might also open up the field of IC design to programmers with little or no hardware background. Two languages which have taken this approach are APECS, described by Boyd in (7) and a variant of the programming language Modula-2, as described by Robinson and Dion (8).

APECS was suggested as a replacement input language for the GAELIC Design Suite, which was used by many UK Universities involved in LSI and VLSI design. APECS designs were to be constructed from 'blocks', which were parts of a design with an identifiable function. Blocks could contain 'components', which were either other blocks or primitives such as transistors. Communciation between blocks and the outside world was permitted through 'ports', and both the external ports of the block and the internal components of the block were inter-connected by 'nets'. APECS designs were broadly divided into two distinct sections. The first described the functional structure and physical implementation of each block in the design, whereas the second section provided utility programs which could manipulate the design, allowing a variety of different formats to be produced which were consistent with one another. For example, production of a circuit schematic for logic simulation might be one format that could be produced, or by taking into account the geometrical information, a slightly different format could be obtained for detailed circuit analysis. Final artwork could be provided either in sticks form for visual inspection, or completely expanded for ultimate fabrication. The following extract is from an APECS design for a Shift Register Cell.

Structural information regarding the circuit is provided by the APECS description, but no information regarding it's function is recorded. The language also suffers from a similar problem to PLAP in that it is far too preoccupied with the physical placement of components, nets and ports.

The notation based on Modula-2, used ideas similar to APECS, but packaged them in a much neater and understandable manner, and in addition, captured a functional description of the design, which could then be used by the associated simulator. Another feature of the language, which distinguishes it from the other previously discussed languages, was that its target implementation was specifically restricted to that of gate-array technology. So far, all the others had assumed the existence of a full-custom layout system. Therefore, because of this restriction, the structure of the overall design system was simplified, as only the problem of placement and routing of the gate-array needed to be solved. A number of other design languages have taken this route, and therefore have provided only semi-custom design facilities to would be users.

To give some idea of the differences between APECS and this notation, another short extract will be given. This time the circuit is a simple equivalence, constructed from four NOR gates. These are structures that are usually found on a gate array.

The above function procedure describes what constitutes a basic equivalence circuit. It gives the defined node a name, and then allocates the required number of NOR gates. Each gate is assigned a name, and the identity of the signal that connects to it. The input and output pins of the node are identified (and numbered for convenience). The

> Defines a basic cell of a shift register array. This cell will invert the input twice under the control of clocks ph1 and ph2.

```
BEGIN
    beginblock('Srcell');
  { Define Component Transistors }
    enh(2,2,1);
    enh(2,4,2);
    dep(8,2,3);
    enh(2,2,3);
    enh(2,4,5);
    dep(8,2,6);

  { Define ports }
    vssport('vssi',1);
    vssport('vssout',2);
    vddport('vddin',3);
    vddport('vddout',4);
    port('ph1in',5);
    port(ph1out',6);
    .
    .
  { Define internal nets }
    vssnet(anon,a);link(0,1);link(0,2);link(2,1);link(5,1);
    vddnet(anon,2);link(0,3);link(0,4);link(3,2);link(6,2);
```

```
net(anon,3);link(0,5);link(0,6);link(1,3);
   .

   .
{ Place Component Transistors }
   place(1,1,12,20);
   turn(2,E);place(2,1,22,24);
   turn(3,ME);place(3,1,29,25);
   .

   .
{ Place Ports }
   placeport(1,METAL,19,0,4,S);
   placeport(2,METAL,19,30,4,N);
   .

   .
{ Place Internal Nets }
   from(0,1);yy(6);newnode(1,1);
   fromnode(1,1);yy(24);newnode(1,2);
   .

   .
{ Place Bounding Rectangle }
   bound(0,0,42,30);
   endblock;
   .

   .
END
```

behavioural description of the circuit is provided by the call of the procedure
Model. The parameter

```
PROCEDURE Equivalence(name:string; a,b,:signal) : signal;
VAR
   alpha,beta,gamma,equiv : signal;
BEGIN
   beginnode('Equivalence',name);
   gamma := Nor2('gamma',a,b);
   alpha := Nor2('alpha',a,gamma);
   beta := Nor2('beta',gamma,b);
   equiv := Nor2('beta',gamma,b);
   OutputPin(0,equiv);
   InputPin(1,a);
   InputPin(2,b);
   Model(EquivalenceModel,NIL);
   endnode;

   RETURN equiv
END;
```

'EquivalenceModel' is a procedure that defines the actions of the circuit and the NIL parameter allows the provision of a 'state' vector, which for this example is not provided as equivalence circuit is purely combinatorial. Finally, the description is terminated and the result, in the form of a signal which stores the result, is returned to the caller.

The functional description is provided in a very similar manner, using the standard Modula-2 language constructs.

```
PROCEDURE EquivalenceModel(pin:cardinal; a:address);
BEGIN
  IF pin[1] = pin[2] THEN
    changepin(0,one,8,10)
  ELSE
    changepin(0,zero,8,10)
  END EquivalenceModel;
```

This description is called by the simulator whenever one of the two inputs to the component change. The state register 'a' is not used by this description, but must, none the less, be declared. As a result of detecting the change of either input, the values on both pins are compared and the pin which is defined to record the result (that is 0, from the call to OutputPin in Equivalence) is changed. The provided procedure changepin alters the value of the pin numbered 0 to either zero or one, taking between 8 and 10 nsecs from the current simulated time.

It should be evident that this description is by far the clearest shown so far, and is the easiest to understand. There are still problems with the continued use of the same identifier in a number of different forms (for example, gamma and 'gamma') and the arbitrary numbering of pins, which is required in addition to their names. Another problem is the rather relaxed way in which the pin parameters to the procedures Equivalence and EquivalenceModel are declared, and how they are of different types in this example. This could be an area of possible confusion to a designer.

Before leaving the area of Programming Language based design languages, mention should be made of work that has been done in the United States of America, in particular by the Department of Defence towards the definition and implementation of a Hardware Description Language to support the Very High Speed Integrated Circuits program (VHSIC), launched in 1980. A number of American companies have collaborated on the project, which has resulted in the definition of the language VHDL, as described by Shahdad et al (9). One major requirement of the project was that the new language be based on the programming language ADA, and in particular, that it should use the appropriate language constructs from ADA, whenever possible.

ADA, historically, had its origin in the Algol-type languages developed in the 1960's and early 1970's. It was, in particular, influenced by the language Pascal. However, its most striking feature was its use of the language construct known as a package. A package had the property that it could contain procedures, functions and datastructures which were to be grouped together logically. What is more, once defined, a package could be instanced several times within the same program, and thus each instance could be in different states of execution. Variables declared within a package retained their values from one invocation to the next, unlike local variables in a procedure or function. It was also possible to ensure that the implementation of a package was kept private from the programs that used it, by requiring that they only used the defined (and published) interface. In this way, an implementor retained the possibility of altering the implementation, whilst maintaining the interface, and knew that all the programs that used the package would not need modification as well. The concept had also been present in other languages, such as Simula (where it was known as a Class) and Modula (where it was known as a Module). This construct permitted the design of Abstract Data Types, whose implementations were modifiable in a safe and sensible manner.

VHDL adapted the package construct and used it in the definition of circuit descriptions. To illustrate some of the powers of the language, a four bit adder description follows.

```
WITH adder_ressource
ENTITY four_bit_adder
  (a,b: IN bit_vector(3. .0);
  cin: IN bit
  cout: OUT bit
  sum: OUT bit_vector(3. .0)) IS
GENERIC
  (delay: TIME := 36ns);
ASSERTION
  delay > 3ns;
  sum'fanout <= max_fanout;
END four_bit_adder;
```

This section completes the definition of the interface to the four bit adder. It specifies what auxiliary definitions are required, and the name of the design entity. It identifies the input and output ports, giving them names and a type. A generic parameter is specified which allows families of devices to be constructed. Different quantities can be parameterised in this way. For example, timing characteristics, size or permitted fanout for an entity may all be specified. This example also specifies the existence of a delay. Finally, assertions can also be supplied. They are facts which must be true at all times.

Once the interface has been provided, the body of the entity is defined. There are two forms, the behavioural and architectural. The behavioural body describes the behaviour using control flow information. It may contain locally declared datastructures and can use the normal language constructs such as loops, assignment and selection. The architectural body defines how the component is structured and decomposed, using data flow descriptions of the architecture. The designer may use either to describe the component, but if it is to contain structural information, then an architectural body must be included.

A behavioural body for the four bit adder might look like this.

```
BEHAVIOURAL BODY adder4b OF four_bit_adder IS
    VARIABLE t, ta, tb: hexsum
BEGIN
    ta := int(a);
    tb := int(b);
    t := ta + tb;
    cout & sum <= bin(t) AFTER delay;
END adder4b;
```

Note the declaration of the local variables t, tb, and ta. The functions 'int' and 'bin' were declared within 'adder_resources'. Observe how the input and output ports to the 'four_bit_adder' are being used within the description.

The architectural body is slightly more involved and introduces the idea of subcomponents.

```
ARCHITECTURAL BODY pure_structure OF four_bit_adder IS
    SIGNAL c: bitvector (3. .0);
    COMPONENT full_adder(cin,i1,i2:IN bit; cout, res:OUT bit);
BEGIN
    FOR i IN 3. .0 GENERATE
      IF i = 0 GENERATE
        full_adder(cin,a(i),b(i),c(i),sum(i));
      END GENERATE;

      IF i > 0 GENERATE
        full_adder(c(i−1),a(i),b(i),c(i),sum(i));
      END GENERATE;
    END GENERATE;

    cout <= c(3);
END pure_structure;
```

The structure defined here is the conventional way of constructing a four bit adder from four one-bit adders. The component 'full_adder' is the one bit adder, and needs to be defined in a very similar manner to 'four_bit_adder' in terms of an interface, a behaviour and an architecture.

9.4 Special purpose design languages

One of the main advantages that was put forward for using programming language based notations, was that to design a working system required a lot less effort than designing a special purpose language. This was certainly true, and provided the designer was occasionally content to put up with restrictions placed upon him by the host programming language, there would be few problems.

However, some researchers felt that such an approach was not really adequate, and purpose built languages (and accompanying design systems) should be constructed. This approach led quite naturally to the idea of a Silicon compiler, a concept discussed in greater detail in Chapter 11.

9.4.1 Semi-custom languages

Designing such languages tended to be much more involved, as syntactic and semantic analysis of source programs was necessary, in addition to the other facilities that were required like layout and simulation. Some projects limited the scope of the problem by once again considering only a particular form of fabrication technology. For example, Model as described by Gray *et al.* (10), is the input language for Chipsmith Silicon Compiler. Others concentrated upon a particular application area. For example, FIRST as documented by Bergmann (11), and developed by the Department of Electrical and Electronic Engineering at the University of Edinburgh. FIRST has been used as a design language to aid in the rapid implementation and investigation of VLSI digital signal processing systems. It accomplishes this by defining networks of pipe-lined, bit-serial operators.

Model was designed for use with a design system that ultimately generated gate arrays. Designs expressed in Model were hierarchical and used the ideas of procedural abstraction, though this time expressed in a specially designed syntax.

For example, consider the requirement to construct a flip-flop. It might be defined in Model as follows.

```
PART inner[p,q,r,s] −> a,abar
   nand[p,q,abar] −> a
   nand[r,s,a] −> abar
END

PART flipflop[reset,clear,clock,d] −> q,qbar
   SIGNAL aout,bout(1:2)
   inner[reset,bout(2) clear,clock] −> −, aout
   inner[aout,clock,clear,d] −> bout
   inner[reset,aout,clear,bout(1)] −> q,qbar
END
```

This example shows the description of two new components, known as PART's in Model, which are the equivalent of procedures in a programming language. The flip-flop is to be constructed from some initially available components called NAND gates. The input and output ports of each component are specified and locally defined SIGNAL's can be declared. The required netlists are defined by calling the appropriate 'part' with the correct parameters.

Having defined the flip-flop, a complete chip would be requested by the addition of the following Model statements.

```
INPUTPAD res,clr,clk,data
OUTPUTPAD q,qbar
flipflop[res,clr,clk,data] −> q,qbar
ENDOFFILE
```

As a result, the correct number of input and output pads would be made available, and suitably connected. The Chipsmith design system is one of the first to be made commercially available.

9.4.2 Full-custom languages

The language STRICT, documented by Campbell *et al.* (12), aims to reduce to a minimum the amount of information that is needed when designing fully custom, integrated circuits. In addition, it attempts to assist the designer, through the use of the design system, to validate the design in a methodical manner. This has been achieved without the need to complicate the syntax of the language unduly. A criticism that might be levelled at the language VHDL, is that being based upon ADA, it must be a large language, which will prove difficult to learn. A similar criticism was levelled at PL/1, and was one of the reasons why the language was not as successful as it might have been, considering its parentage and what it set out to achieve. As in VHDL, STRICT requires that an interface be defined, along with a behavioural description and ultimately an implementation. To try to achieve the goal of a simple syntax, an applicative or functional style of language was chosen.

The STRICT description for a Shift Register pair is shown.

This description is in two parts. First there is the request for an instance of a particular 'block'. This corresponds, once again, to the call of a procedure, and is followed by the declaration of the necessary blocks to fulfil the request. This request also arranges for the correct number of input and output pads to be made available, including power and ground and that they are suitably connected. It allocates names to the pads so that checkplots or similar pictorial representations will be suitably annotated.

There then follows the second part of the description. The delaration of the block is also split into two distinct sections, of which the first must always exist. The first contains the definition of the interface plus a behavioural description while the second contains the implementation, if it exists. Blocks can support, if necessary, generic parameters, assertions, new type definitions, inheritance from libraries and elsewhere of blocks or types, and conditional implementations, controlled by the use of guards. Pragmats to the design system are permitted to allow relative placement of ports around the boundary of the block,

```
BUILD
  {INSTANCE
    pair:srpair(10,10)
  USING
    pair(pairin,clock1,clock2)
  MAKE
    pairout ::= pair.output
  }
GIVEN
  BLOCK srpair (rise, fall : INTEGER)
    HAVING (input@W : WIRE
            clock1@N, clock2@S : CLOCK(rise,fall):
            (output@E : WIRE)
  INTENDED BEHAVIOUR
    WHENEVER
      rise(clock1):
        WITHIN cycle(clock1)
          SET
            output = input
  USE STRUCTURE
  {INHERIT
    srcell
  INSTANCE
    si,sj : srcell
  PLACE
    si;sj
  USING
    si(input,clock1)
    sj(si.output,clock2)
  MAKE
    output : := sj.output
  }
END
```

and relative placement of blocks with respect to each other. The above example illustrates some of these features.

At the present time the front end of the STRICT design system is completed plus a substantial part of the layout system. Integration of the simulation system continues. It is proposed that in the future, a verification system will also be developed which would allow formally based verification rules to be applied to STRICT designs.

9.5 Conclusions

It has been shown that the use of high level design languages of increasing versatility is increasing. What is more, the increasing power of the languages will permit designs of many millions of gates to be attempted in the knowledge that many of the practical complexities are being handled automatically, allowing the designer to concentrate on getting the design correct.

It is also worth noting that languages are also being developed not for use in the actual design process, but for the interchange of design information from site to site. EDIF as described in (13) is a current proposal for such a language. It has arisen in the current climate of increasing collaboration between industry and academia, and even wholly within industry itself. In the past, languages that were used for exchanging data tended to be very special purpose, and often company confidential. They were very difficult to implement well, and even more difficult to extend as the technology progressed. EDIF has therefore been designed as a common interchange format, available to all, with a well defined syntax and semantics. It is currently aimed at the gate-array and semi-custom design markets, as this is where the demand is greatest at the current time. It is not intended as a design language, but rather as format for the transmission of design data.

The February (1985) Issue of 'IEEE Computer' (14) devoted itself to the whole question of Design Languages for Silicon.

The future heralds a move towards the formal verification of designs and more automatic generation of layout. Verification is covered in more detail in Chapter 14, whilst some aspects of Automatic Layout were covered in Chapter 6. In a sense, it will correspond to the way in which Software Programming Languages developed after the experiences gained with the first FORTRAN compilers had been analysed. However, this time the developments will be a lot quicker as our expertise in language design is now quite considerable. And with these new skills, the complexities of VLSI Systems Design will, it is hoped, be mastered.

References

1   MEAD, C., CONWAY, L.: 'Introduction to VLSI Systems', 1980, Addison-Wesley.
2   MAVOR, J., JACK, M., DENYER, P.: 'Introduction to MOS LSI Design', 1983, Addison-Wesley
3   JOHANSSEN, D.: 'Bristle Blocks: A Silicon Compiler', 1979, Proceedings, 16th Design Automation Conference

4 LOCANTHIS, B.: 'LAP: a SIMULA package for IC layout', 1978, Caltech Display File No. 1862

5 ANON.: 'VLSI Design Tools', Internal Report, Dept. of Computer Science, University of Edinburgh

6 FLETCHER, M. D., MOLE, G. F.: 'VLSI Design Tools Manual", Internal Doc., Dept. of Electrical and Electronic Engineering, 1983, University of Newcastle upon Tyne

7 BOYD, D. R. S.: 'APECS: A Pascal Environment for Circuit Specification: A Preliminary Report', 1981, Technology Division, Rutherford and Appleton Laboratories

8 ROBINSON, P., DION, J.: 'Programming Languages for Hardware Description', 1983, 20th Design Automation Conference

9 SHAHDAD, M., LIPSETT, R., MARSHNER, E., SHEEHAN, K., COHEN, H., WAXMAN, R., ACKLEY, D.: 'VHSIC Hardware Description Language', February 1985, IEEE Computer

10 GRAY, J. P., BUCHANAN, I., ROBERTSON, P. S.: 'Controlling VLSI Complexity using a High-Level Language for Design Description', 1983, 20th Design Automation Conference

11 BERGMANN, N.: 'A Case Study of the FIRST Silicon Compiler', Dept. of Computer Science, 1984, University of Edinburgh, Internal Report, CSR-159-84

12 CAMPBELL, R. H., KOELMANS, A. M., MCLAUCHLAN, M. R.: 'STRICT', IEE Proceeding, March/April 1985, Vol. 132, Pts E. and I., No. 2

13 ANON.: 'EDIF: Electronic Design Interchange Format, 1985, Version 1 0 0 Specification

14 IEEE Computer, February, 1985

Chapter 10

PLA design tools

E. G. Chester

10.1 Introduction

The use of CAD tools for designing digital integrated circuits inevitably involves some form of trade-off between regularity and flexibility of the structures used. Highly regular structures such as the gate array, aside from their low manufacturing costs for small volume production runs, lend themselves well to the use of CAD tools since the limited range of possible structures makes it possible to characterise and model these structures extensively.

At the other end of the spectrum there is full custom design which allows the designer more degrees of freedom to explore in the search for the optimum solution to a design problem. This expansion of the solution space of the problem requires a large increase in the complexity and power of computer tools to aid the designer, or more likely, a reduction in the proportion of the task which can be automated.

Some designers have to work at the leading edge of current technology to achieve the very high speed and performance needed by specialised applications. Other applications require only a moderate performance from the devices, but they must be simple and cheap to design and prototype. The gate array and standard cell approaches lie towards this end of the spectrum.

Between these extremes lies a range of applications which require a full mask set for fabrication of specialised structures such as memory arrays and computation elements along with interconnection and control logic. Much of the logic in this type of design can be implemented using the programmable logic array (PLA), restricting manual intervention to cells which are in the critical path. The PLA can be incorporated into a design system as a parameterised macro cell which, once created, can be placed and routed using the same tools as custom and library cells.

The ability to 'mix and match' several design styles on the same chip is a major advantage of full custom fabrication processes and the PLA is an important component of this technique.

10.1.1 Limitations of regular structures

It is a feature of current VLSI technology that interconnections tend to dominate the size of a layout. Irregular interconnection schemes tend to expand layouts since changes of direction and inter-layer contacts increase the area required for wiring. The PLA represents an attempt to regularise the interconnection scheme, producing a structure which can compare favourably in size with irregularly connected random logic. This technique is referred to by Ayres (1) as 'crystalline' layout, since it represents an optimal packing of small structures in a regular arrangement, analogous to the packing of molecules in a crystal which produces a minimum of potential energy. As the size of a PLA is expanded, there comes a point at which the efficiency deteriorates. This happens for two reasons. Firstly, a large PLA contains long signal and power lines which can lead to excessive signal delays and a large power supply source resistance. These effects can be offset by introducing extra buffers and power connections at intervals in the array. Secondly, the functional complexity of a large PLA may exceed the connectivity limitations of the array. The point at which this occurs varies depending on the nature of the logic function and how well it is matched to the connectivity of the array.

10.1.2 The 'Designer's toolbox' approach

A useful way of integrating the PLA with other design aids is the 'Designer's Toolbox' approach promoted at Berkeley and exemplified by Newton *et al.* (2). The tools are implemented as a collection of programs which are held together by a common operating system environment. Data is exchanged between programs in text files which can be interpreted easily by the user or by a program written in a high level language. This allows a moderately experienced user to write a specialist tool which will slot into the system without the problems of creating an input parser, database manager etc.

Using this type of system, the flexibility of the PLA may be utilised to the full by the provision of a set of tools which perform bit map production, logic minimisation, folding, simulation, test generation and conversion of various forms of functional specification.

10.2 Structural variants

The PLA is not a single structure, but rather a class of structures including many different forms. These variations each have their own special requirements and hence a variety of design tools may be needed.

10.2.1 Combinational functions

10.2.1.1 The basic PLA: In its most basic form, the PLA directly implements a two-level logic function using a set of input true/complement buffers followed by a level of AND gates then a level of OR gates, or their equivalents. This is described by Fleisher and Maissel (3). As stated in 10.1.1 there is a connectivity limitation in this simple array. This is a serious problem in functions which make extensive use of the exclusive-or operation, such as parity and addition. Some considerable improvement is possible by introducing an extra level of logic, as illustrated in Russell *et al.* (4).

10.2.1.2 The three-level PLA: An extra level may be added to the PLA in a general way by adding an extra logic plane to produce (e.g.) an AND-OR-AND structure. Many functions, however, require that the exclusive-or operation be performed only between input signal pairs. In these cases, it is more efficient to place two-bit decoders on adjacent input pairs in place of the true/complement buffers.

This scheme was extended by Weinberger (5) to include exclusive-or gates between adjacent PLA outputs, allowing a complete adder to be produced in one PLA without an exponential growth in size with the number of bits – a problem which quickly renders the basic PLA useless in this application. This scheme was further extended by Schmookler (6) to produce a complete ALU incorporating Boolean, arithmetic and flag operations and a carry look-ahead unit for extending the number of bits even further.

10.2.1.3 Masking operations: Some functions, such as data path functions, require operations to be performed on one of a set of several disjoint subsets of inputs. This can be performed in a PLA, but is inefficient and sparse in its use of the array. An improvement can be obtained as shown by Jones (7) by selectively degating subsets of inputs when they are not required, i.e. incorporating some selection logic into the input buffers which forces the array inputs to the 'don't care' state. (Fig. 10.1). This can also be applied to the output buffers where required.

It is up to the designer to decide whether the added complexity of the buffers is justified, or whether the bus switching should be performed external to the PLA. This technique has some applications in enhancing testability, however.

10.2.2 Incorporation of memory into the PLA

10.2.2.1 The finite state machine: The most frequently encountered use of memory in a PLA is the use of a latch to feed back outputs to inputs to form a synchronous FSM. This allows sequence and control logic to be easily implemented, as described by Logue et al (8) and more recently by Mead & Conway (9).

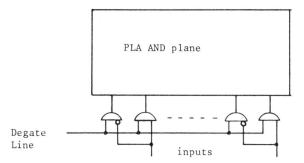

Fig. 10.1 *Degating applied to PLA input buffer*

10.2.2.2 The writeable PLA: Instead of hard-wiring the personalisation data into a PLA, the crosspoints can each have a memory bit to enable or disable them as described by Fleisher and Maissel (3). This can be considered as a PLA whose function can be changed as required, or alternatively as an associative or content-addressable memory. (Fig. 10.2).

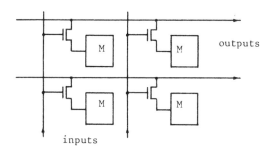

Fig. 10.2 *Writeable PLA Cell*

10.2.2.3 The storage/logic array: The SLA, described by Patil and Welch (10) is a PLA in which the input and output columns are interleaved, thus merging the AND plane with the OR plane. Memory cells are inserted in rows at regular intervals between the rows of cross-points, allowing outputs to be fed back to inputs.

The advantages of the SLA are that it is a highly flexible way to implement sequential logic systems and that it is possible for CAD tools to map a complete design comprising combinational and sequential subsystems onto one SLA in a regular way. The disadvantages are that the size of the SLA may not be optimal due to the lower density of crosspoints compared to the PLA and the testability problems which may arise from the distributed nature of the state information.

10.2.3 Circuit level implementation

10.2.3.1 The static PLA: It is simpler to employ static logic gates where possible. This is straightforward using NMOS technology, providing that NOR logic is used to avoid the long series chain of transistors required for NAND logic. (The delay of such a chain increases as the square of its length, since both resistance and capacitance are distributed along it.) (Fig. 10.3a).

Using CMOS technology, the requirement to implement switch paths for both a function and its dual makes the use of a series chain unavoidable and doubles the number of transistors required in the array, unless pseudo-NMOS logic is used. (Fig. 10.3b).

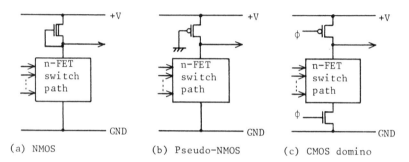

(a) NMOS   (b) Pseudo-NMOS   (c) CMOS domino

Fig. 10.3   *Gate types for PLA AND/OR planes*

10.2.3.2 The dynamic CMOS PLA. This form of implementation reduces the static power consumption of the PLA but requires some care to avoid charge sharing problems. Since domino logic (Fig. 10.3c) requires clocking, some extra clock phases may be required to prevent possible hazards when connecting such a PLA to clocked latches to produce a FSM.

10.3 Computer tools

10.3.1 Input description processing

The function of a PLA can be expressed in a variety of forms depending on the application. It may be in the form of Boolean logic expressions, perhaps with some bracketing of terms and higher level functions such as exclusive-or. It may be a table of cubes in which each input or output is listed as 0, 1 or X. In the case of a Weinberger adder PLA, the AND plane entries may be specified as the symbols for the propagate, generate or half-sum functions of the inputs (or their complements).

In all of these cases, the input description must be converted to the form of a bit map for folding and mask layout. Some tools such as test pattern generators and logic minimisers may operate more efficiently on a standardised cube notation.

10.3.2 Layout generation

A tool can easily be created to lay out the required size of array from layout macro cells. The function of the PLA is then programmed by overlaying crosspoint connections according to the bit map description file. The layout tool may be required to automatically adjust the size of buffers and wires or insert extra ones according to the size of the PLA.

It may also be required to generate different structural variants according to some input parameters, such as the inclusion of input decoders or storage elements.

Another useful feature is the ability to vary the positions of the inputs and outputs as required by topological minimisation or folding.

10.3.3 Finite state machine compilation

An example of a tool of this type is shown in Russell *et al.* (4). The description of a state machine by the designer is often in the form of a state transition diagram or an algorithmic state machine (ASM) chart. This can be described in the form of a text file and fed to a FSM compiler which can assign state bits and translate the description to a bit map for a PLA with feedback.

Some optimisation may be possible in the way in which the states are encoded in order to minimise the resulting PLA size.

10.3.4 Simulation and verification

Simulation of the PLA can usually be carried out at a fairly high level, since the regular structure allows timing and loading parameters to be easily predicted once an appropriate model has been produced. A simulator can read in the bit map for a PLA in order to simulate its logic function, ensuring consistency between simulation and layout, since both derive from the same data.

A method favoured by the author is the use of a set of register transfer level simulation procedure embedded in a high level programming language. This permits some degree of verification to be achieved. An example of this is the design of a mod 31 adder which was verified in this way. Simple enumeration of the full range of possible inputs produced a large quantity of data which was checked by comparison with the same function generated by the arithmetic of the programming language. Such an approach is not as thorough as formal logic proof, nor is it very efficient computationally, but it does employ the computer in the type of task which it does best, preventing the error-prone process of checking large quantities of simulation output by inspection. There is also some degree of confidence that if two different implementations of the same function match each other, they probably match the specification.

10.3.5 Logic minimisation

Since it is desirable to produce a PLA which is as small as possible, logic minimisation is performed to achieve this. There are three aims in this reduction according to Kambayashi (11). The first is to minimise the number of inputs by elimination of any which are redundant. Secondly, the outputs may be produced in complementary form if this reduces the number of product terms. Finally, the total number of product terms must be minimised.

If a classical approach is used to perform the product term reduction such as the Quine and McCluskey method, the number of prime implicants which must be generated grows exponentially with the number of inputs. The problem can be attacked more effectively by using a heuristic method such as that employed in the MINI system described by Hong *et al.* (12). The function is encoded in a cubical notation in two cover lists; one for output one values, one for output 'don't care' values. An iteration then takes place in which the cubes are expanded into the 'don't care' region, reshaped and merged, then reduced by removal of redundant cubes and trimming of the remaining ones. These steps are reiterated until no significant reduction is observed between steps.

This process may not find the true minimum solution, but it can approach this very closely within a relatively short time.

10.3.6 Folding
Once logic minimisation has been carried out, some further size reduction may be possible by folding the PLA. This involves the sharing of columns and rows by more than one signal, hence making it necessary to route connections from the array on three of four sides. According to Egan and Liu (13), arbitrary Boolean functions produce sparse PLAs in which typically 90% of the crosspoints are unused. Folding achieves size reduction by removal of areas of unused crosspoints and compaction of those remaining.

Tools for performing this process are included in the set of tools developed at Berkeley described by Newton *et al.* (2), namely BLAM and PLAFOLD.

10.3.6.1 Simple folding: This folding technique consists of simple column folding which requires inputs and outputs to be connected at both ends of a column, and simple row folding which requires the AND plane or OR plane to be split up and product terms to be available at each side of the plane.

10.3.6.2 Optimal folding: The problem of finding a minimum size folding appears to be NP-complete (Hachtel *et al.* (14)). When a cut is made in a column to allow two columns to be merged, this constrains certain rows to be positioned above or below the cut. This may constrain the subsequent folding operations and block further progress. The method employed by Hachtel *et al.* (14) is to use a heuristic algorithm based on a graph theoretic description of the problem. The reduction in area obtainable is quoted as typically 30% to 40%.

10.3.6.3 Bipartite folding: Egan and Liu (13) propose a reduction of the problem complexity by considering only bipartite folding, where all cuts are made in the same row or column. They employ a branch-and-bound algorithm which is claimed to achieve results close to the optimum. An extra feature of this algorithm is that it can be used to partition large PLAs into smaller ones.

If input and output columns are folded together, a bipartite folding can produce interlocking L-shaped AND and OR planes.

10.3.6.4 Multiple folding: When an extra routing layer such as 2nd level metal or polysilicon is available, it is possible to make more than one cut in each column. This requires the 'isolated' section of the column to be connected orthogonal to the column on the extra routing layer. A folder which can operate in this way is the PLEASURE program (De Micheli and Sangiovanni-Vincentelli (15)).

This program also attacks the problem often encountered with folding, namely the changes in the placement of input and output connections which can adversely affect the routing of tracks to the PLA. The PLEASURE program allows the designer to specify constraints on the placement of connections which prevent the program from changing them. This reduces the level of optimisation which can be achieved, of course.

10.3.7 Test generation

The automatic production of test patterns for a PLA can be performed by considering the structure as a set of AND and OR gates and applying a standard method such as path sensitisation. The difficulty which arises is the rapid increase in the size of the computation with the number of gates, especially since the gates can have many inputs.

To avoid the lengthy computation required by deterministic methods, the PLA/TG program described by Eichelberger and Lindbloom (16) uses a heuristic method. The algorithm attempts to generate test vectors for each product term such that:

(i)   All the inputs to the AND gate generating the term are at logic 1. (Test input stuck-at-0, product term stuck-at-0)

(ii) One input to the AND gate is at logic 0, the remainder are at logic 1. (Test input stuck-at-1, product term stuck-at-1)

The path through the OR plane must be sensitised by assigning the 'don't care' inputs for the product term such that only the desired product term is active. The approach used by PLA/TG is to randomly assign the 'don't care' inputs since there is a high probability of generating a type (ii) test which helps sensitise the required type (i) test. The type (i) and (ii) tests can be subsumed to produce a single test vector, but type (i) tests will block OR-plane sensitisation if they are subsumed.

The process is repeated until a test set is achieved for each product term. If several passes fail to generate any more fault cover, subsumption is dropped to prevent patterns from deadlocking and mop up remaining faults. Finally, any uncovered faults are detected by a closing routine which randomly tries all possible 'don't care' assignments. (This is usually needed when the number of product terms is high, the number of used AND crosspoints is low and many product terms feed an OR gate.)

The PLA structure can exhibit non-classical faults since it is possible for extra crosspoints to occur where they are not required. These can be modelled by the introduction of extra inputs to the gates in the model, with a subsequent increase in the length of the test generation process. Some test generation programs such as PART described by Somenzi *et al.* (17) and TPLA described by Ostapko and Hong (18) are designed specifically to consider crosspoint fault models.

A convenient bonus provided by PLA/TG is the flagging of redundant crosspoints, which may indicate a design error.

10.4 PLA Testability

The usefulness of the PLA has led to a great deal of interest in the special problems of testing systems which contain PLAs, since the ease of design and layout are of little use if the resulting device is very difficult to test.

10.4.1 Enhanced PLA controllability and observability
Once a set of test vectors had been generated, it must be possible to apply the vectors to a PLA embedded in a circuit. One well known method is the scan path, which can be built into a PLA by replacing the input and output buffers by shift register latches.

The drawback of this method, apart from the increase in size, is that long shift registers, such as will be needed for scan path on VLSI devices, can lead to very long testing times for clocking sequences in and out.

10.4.2 Self-testing PLAs
Daehn and Mucha (19) propose a scheme in which a BILBO register is used at the input and output of the PLA, and also between the AND and OR planes. This can produce random test patterns and produce a signature analysis of the output at the full operating speed of the device. Unfortunately, the large fan-in of the gates used in PLAs makes the probability of producing a useful test with a random pattern quite low, requiring the BILBO to be clocked many times, producing exhaustive testing in the limit.

Modified PLAs have been proposed by various workers in the field. For example, Grassl and Pfleiderer (20) have developed a PLA which incorporates a test generator at the input which produces the all-ones state at the input of the AND place then shifts a single zero through every input position, generating a full AND test set (Fig. 10.4). To sensitise only one product term at a time, a second shifter is used to mask off all the product terms except one.

The mask is then shifted to enable each product term row in turn. The output patterns are accumulated in a signature register built into the output buffers for later examination.

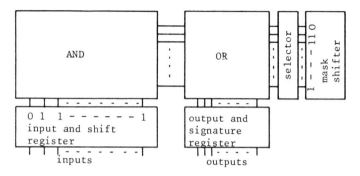

Fig. 10.4   *Self-testing PLA*

A random pattern testable PLA is described by Eichelberger and Lindloom (21) which relies on the fact that small PLAs are more easily tested by random patterns (Fig. 10.5). The PLA contains degating logic on the inputs and product terms which can partition the PLA into small sections, only one of which is selected. The selection is done by decoding logic which is driven by outputs from the random pattern generator.

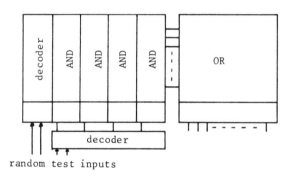

Fig. 10.5   *Random pattern testable PLA*

10.4.3 Code checked PLAs

If a PLA is designed only to generate valid code words from a code such as the k-out-of-n code, a checker can be used to verify that the code word is valid and signal faults to the outside world. The checker itself can also be implemented as a PLA, as shown by Bose and Lin (22).

The large increase in size caused by this method can be reduced by using a low-cost residue code (Sayers and Kinniment (23)).

10.5 Conclusion

The PLA has been shown to be a versatile logic building block which can be supported by a wide range of computer tools. The ease of use leads designers to use the PLA in preference to other forms of logic which are marginally more efficient, but require more design effort. Some applications which are not directly suited to the PLA can be catered for by specially designed PLAs.

The PLA is not a panacea, however, and the designer must decide whether the circuit performance requirements permit it to be used. If the tools available allow the PLA to be closely integrated with other layout macros and custom-designed elements, then maximum flexibility to the designer is assured.

References

1 AYRES, R. F.: 'VLSI: Silicon compilation and the art of automatic microchip design', 1983, Prentice Hall Inc., Englewood Cliffs, New Jersey
2 NEWTON, A. R., PEDERSON, D. O., SANGIOVANNI-VINCENTELLI, A. L. and SEQUIN, C. H., *IEEE Trans. Circuits Syst.*, CAS-28, 1981, 666–680
3 FLEISHER, H. and MAISSEL, L. I., *IBM J. Res. Develop.*, *19*, 1975, 98–109
4 RUSSELL, G., KINNIMENT, D. J., CHESTER, E. G. and McLAUCHLAN, M. R.: 'CAD for VLSI', Van Nostrand Reinhold (UK), Wokingham, England, 1985, Chapter 5
5 WEINBERGER, A., *IBM J. Res. Develop.*, *23*, 1979, 163–178
6 SCHMOOKLER, M. S., *IBM J. Res. Develop.*, *24*, 1980, 2–14
7 JONES, J. W., *IBM J. Res. Develop.*, *19*, 1975, 120–126
8 LOGUE, J. C., BRICKMAN, N. F., HOWLEY, F., JONES, J. W. and WU, W. W., *IBM J. Res. Develop, 19,* 1975, 110–119
9 MEAD, C. A. and CONWAY, L., 'Introduction to VLSI Systems', Addison-Wesley, Philippines, 1980
10 PATIL, S. S. and WELCH, T. A., *IEEE Trans. Comput.*, C-28, 1979, 594–601
11 KAMBAYASHI, Y., *IEEE Trans. Comput.*, C-28, 1979, 609–617
12 HONG, S. J., CAIN, R. G. and OSTAPKO, D. L., *IBM J. Res. Develop.*, *18,* 1974, 443–458
13 EGAN, J. R. and LIU, C. L., *IEEE Trans. Computer-Aided Design*, CAD-3, 1984, 191–199
14 HACHTEL, G. D., NEWTON, A. R. and SANGIOVANNI-VINCENTELLI, A. L., *IEEE Trans. Computer-Aided Design of Integrated Circuits and Syst.*, CAD-1, 1982, 63–77
15 De MICHELI, G. and SANGIOVANNI-VINCENTELLI, A. L., *IEEE Trans. Computer-Aided Design*, CAD-2, 1983, 151–167
16 EICHELBERGER, E. B. and LINDBLOOM, E., *IBM J. Res. Develop.*, *24*, 1980, 15–22
17 SOMENZI, F., GAI, S., MEZZALAMA, M. and PRINETTO, P., *IEEE Trans. Computer-Aided Design*, CAD-3, 1984, 142–149
18 OSTAPKO, D. L. and HONG, S. J., *IEEE Trans. Comput.*, C-28, 1979, 617–627
19 DAEHN, W. and MUCHA, J., IEEE Trans. Comput., C-30, 1981, 829–833
20 GRASSL, G. and PFLEIDERER, H.-J., *INTEGRATION, 1*, 1983, 71–80
21 EICHELBERGER, E. B., and LINDBLOOM, E., *IBM J. Res. Develop.*, *27*, 1983, 265–272
22 BOSE, B. and LIN, D. J., *IEEE Trans. Comput.*, C-33, 1984, 583–588
23 SAYERS, I. L. and KINNIMENT, D. J., *IEE Proceedings Pt. E, 132*, 1985, 197–202

Silicon compilation – design and synthesis beyond CAD

S. G. Smith

11.1 Introduction

Recent years have seen relentless advances in IC fabrication technology. It is now possible to produce devices containing hundreds of thousands of transistors and exhibiting formidable functional complexity. The expertise required to design such VLSI devices ranges from circuit theory, through logic design, to hardware and software architecture. Product lifecycles are becoming shorter while demand for new designs is on the rise. These circumstances have conspired to produce an acute shortage of IC designers.

To meet the challenge of ever-increasing design complexity [1], the creators of VLSI devices have turned to the computer for help. While CAD software tools for layout, circuit-extraction, design-rule checking and simulation assist the designer in his task, the time and effort required to bring a new IC into being remains unacceptably high. Often a design is found to be defective on fabrication, necessitating redesign cycles which are expensive both in terms of processing cost and lost market opportunity.

An attempt was made to increase the IC designer base by simplifying the engineering aspects of circuit design, bringing structured IC design to the undergraduate student [2]. Alternative semi-custom approaches, such as gate arrays and libraries of standard cells [3] allow an initial design investment to be carried over into subsequent designs. However while the problem of complexity remains, the designer cannot be sure of correctness, and product lead times are too long to keep pace with available technology.

The silicon compiler [4,5] is proposed as a solution to the complexity problem, enabling the user to say what he wants done without having to do it all himself, and bringing custom IC design to the wide community of systems engineers. In a sense this conflicts with mainstream CAD, in that it displaces skilled IC designers instead of assisting them.

11.2 What is a silicon compiler?

Silicon compilers, first described in 1979 by Johannsen [6], and forming the central theme of his Caltech thesis, are an attempt to 'bridle' the complexity problem by abstracting the details of the design to the highest levels of conception. They make use of a 'knowledge base' of layout, timing and simulation strategies, allowing these low-level implementation details to be effectively hidden from the designer, which frees him to work at the architectural and functional level, where arguably his talents are put to best use. After this knowledge base is verified, all designs which can be described in the hardware description language (HDL) will be realisable and guaranteed to work – this is known as correctness by construction. Behavioural or structural specifications are translated into correct chip layout, the translation process being completely automatic.

11.3 Silicon and software compilers

A similar transformation took place some years ago in the field of computer programming. As the power and memory size of computers grew, programmers could produce larger and more complex programs, whose design and debug times increased accordingly. The solution to that complexity problem was the high-level compiler, which today is used for practically all programming. The slight reduction in efficiency of resulting code is more than compensated for by the enormous reduction in design effort. Furthermore, there is no need to verify that the machine code correctly implements the source specification – the compiler, being known good, always produces code that is correct by construction. Any malfunction of the object code must result from errors of specification originating from the designer.

The software programmer, writing in a high-level language, is able to describe the desired program at a behavioural level, and can easily extract information from his high-level source file. This is infinitely preferable to working with a long list of hex codes. He often does not need to know the exact physical addresses in which his code will reside, how the compiler allocates registers, nor even the instruction set of the target machine. In some cases, even the identity of the target machine is of no interest. However critical code (e.g. inner loop computation) may be written in machine-specific assembly language, and loaded at compile-time.

Should he inadvertently enter illegal code, the compiler will flag an error or take some corrective action of its own. The inevitable specificational changes which occur during the course of a project need not lead to massive redesign costs – the programmer is able, using his favourite editor, to make rapid modifications to his design file, and recompile the program. Revisions and their documentation are easy, any effort required being at high level.

11.3.1 Similarities

Silicon compilers offer the same advantages to the digital systems designer as do software compilers to the programmer. The designer is able to specify his system by intent, rather than as a mass of arcane geometrical shapes, and can far more easily keep pace with progress in his design. Low-level detail such as mask layers used, exact physical location of functional elements, and design rules are hidden from him. However should performance of any module be critical, it can be designed by hand (using traditional tools) and incorporated into the compiled chip.

Erroneous code which would lead to such undesirable phenomena as floating or undriven nodes, is flagged by the silicon compiler. Correctness by construction guarantees working silicon, eliminating the need for design-rule checking (DRC), whose computational cost in VLSI is astronomical. Implementation and documentation of revisions is a simple matter.

11.3.2 Differences

Some differences exist between the two compiler types. The output of a software compile is a list of code, having in effect one dimension – that of length. Any object in the list is identified by one number, its address. The silicon compiler produces a two-dimensional pattern of mask geometries, and any device is identified by two numbers – an x- and a y-coordinate. It follows that the number of placement options (as a function of device count) grows rapidly.

A second difference is in the cost of communication. A software jump is easy to implement, and the cost is independent of distance. The cost of communication on an IC rises with the distance, in terms of area, time and power consumption. Furthermore, the occasion might arise where obstacles make communication impossible. A software jump does not suffer from this problem.

Both the dimensionality and the communication problems can be solved through the use of hierarchy. Silicon compilers make use of the specificational hierarchy, equating functional and physical hierarchy as directed by the floorplan (a blueprint or template used by the compiler to construct chips, thereby imposing necessary architectural restrictions). Communication between modules can only occur in well-defined manner at equal or adjacent levels in the hierarchy, and, as the floorplan guarantees these two types of communication, any chip describable in the language can be compiled.

11.3.3 Design criteria

Once upon a time software programmers had to extract every last chunk of performance from the target machine. Now they are more concerned with software compatibility, documentation, testability etc. Language compilers have removed the low-level worries, leaving them clear to work at the important higher levels.

As well as technical advances, improved interfaces between VLSI design and fabrication communities (in the form of silicon foundries, multi-project wafers etc., e.g. MOSIS [7]) have caused a drop in the price of silicon processing, fuelling demand for new designs and boosting the application-specific IC (ASIC) market. Thus we see similar changes currently taking place in the field of VLSI design – issues such as design confidence, productivity and testability are beginning to take priority over yesterday's optimisation criteria of yield and performance.

As the art of software compiler design reaches maturity, it is claimed that the new optimising compilers produce code more efficient than that produced 'by hand'. Similarly it is possible that later generations of silicon compiler may produce denser IC designs than those handpacked by experts.

11.4 Design leverage and the complexity pyramid

The way in which silicon compilers allow ease of correct design is by hiding details from the designer below a certain level. The human mind can only keep track of a finite number of objects at once. In the early days of IC design, the number of transistors on a chip was within the designer's management capability. Hierarchical design specification was unnecessary. As complexity increased, approaches such as standard cell or gate arrays allowed the hiding of the lowest levels of implementation from the designer. Through the use of 'macros', further details could be hidden. Even these techniques are obsolescent, as VLSI complexity grows further and further into the systems area.

Such techniques are analogous to building a 'pyramid' of levels of abstraction, and freeing the designer to work in the higher portions of the pyramid, above a 'leverage point'. Maximum design leverage is attained by holding design activity to the highest level of abstraction. VLSI complexity is managed by the automatic synthesis of all views of the circuit below the leverage point from a description at architectural or functional level. The effect of one simple modification at high level grows as it propagates down the pyramid. At transistor level, an equivalent modification would require man-months of effort, with the attendant risk of introducing errors.

11.5 Restrictive practices

The leverage just described can only be achieved at the cost of designer freedom. The same simplifications and abstractions which allow novices to design LSI devices [2] can be used to automate the realisation of VLSI devices. This can be viewed as both a good thing and a bad thing – while a naive designer benefits from working in an environment which guarantees correctness, an experienced designer can experience frustration in not being able to employ his favourite tricks and shortcuts. Devices produced by current silicon compilers are not yet comparable in terms of area, speed, power consumption etc. with those hand-crafted by experts. However the productivity offered by silicon compilers is undoubtedly a boon to experienced and inexperienced designers alike. What is lost in yield and performance is more than compensated by reduced design costs (man hours, computer time), and opportunity costs (time spent getting the product to market).

Design for testability [8] may also be enhanced by designer restriction. If designer input is maintained above the circuit level, then gates and latches which are reconfigurable for test and/or self-test may be included by the compiler, unseen by the designer. From then on, automatic test-pattern generation (ATPG) is a relatively simple matter.

11.6 A review of the field

Some new companies have recently come to market with silicon compilers. Many of their projects began life on a University campus (the 'R' part), and evolved into a product (the 'D' part) when transferred into industry. This is by no means a complete review – there are countless silicon compiler projects in CS departments, industrial and government research centres worldwide which have yet to bear fruit. We begin with the compiler best known to the author – FIRST, for Fast Implementation of Real-time Signal Transforms.

11.6.1 FIRST (University of Edinburgh)
FIRST [5] emerged as a joint project between the Departments of EE and CS at Edinburgh in 1982. Restricting its application to a specific class of problems (signal processing), and using bit-serial computation and communication, the designers of FIRST were able to come up with a very efficient layout strategy. Although essentially technology-independent, FIRST has been built around a 5-μm nMOS cell library, and a two-phase non-overlapping clock scheme. A 2.5-μm double-metal CMOS library is under development.

The user is offered a single, high-level interface to FIRST in the form of a network description language. FIRST has (unseen by the user) a library of hand-designed 'leaf-cells', and software procedures for their assembly into parameterised 'primitives' (multiply, add etc.). Primitives form the lowest level of hierarchy available to the user, and may be used to construct flow-graphs of higher level objects (operators, chips, subsystems and finally systems) to solve specific DSP problems. An analogy to software would cast primitives in the role of 'instructions', while the other hierarchical elements are like user-defined 'sub-routines'. The first language and library is user-expandable, in that the user may design and append new primitives according to well-defined guidelines.

Chips are assembled according to a simple layout procedure, where two ranks of bit-serial processesors communicate via a central channel. Pads are sited in an external rectangular 'ring', with space for optional linear feedback shift registers and control circuitry for purposes of self-test [9]. Serial communication allows most of the chip area to be devoted to computation, and routing of the central channel is a relatively simple problem.

No silicon compiler allows the specification of low-level design errors (e.g. timing, race-conditions, design-rule violation, etc.). FIRST has the facility to warn the user when he strays from the path of design integrity in an obvious manner (e.g. nodes overdriven or left floating, word-synchronisation errors etc.). However high-level errors may creep in, which cannot be flagged by either language compiler or placement software (e.g. inputs to a subtractor reversed). Each primitive has a behavioural model which is exercised by a word-level, event-driven simulator for functional verification throughout the design process. It is crucial that behavioural duality is maintained between hardware and software models – this is ensured from the user's point of view by driving the simulator and layout software from the same source file.

FIRST guarantees not only functional correctness but also performance, due to the low pin/buffer count benefits of serial communication. Each cell in the library is designed to work at a fixed, process-dependent clock-rate. The floorplan guarantees data transfers both between primitives and between chips at this rate. A recent comparison study [10] between FIRST and commodity 'micro-DSP' implementations of a large signal processor showed impressive savings in design time, board size, power consumption, chip and pin count.

11.6.2 Chipsmith (Lattice Logic Ltd.)

Lattice Logic was formed in 1982 by John Gray (who had previously directed the Silicon Structures Project at Caltech), and Irene Buchanan, both of whom are ex-Edinburgh University CS department (and now departed to the recent ASIC start-up company, European Silicon Structures). It is no coincidence that two prominent silicon compilers were developed in the same city – Gray and Buchanan influenced the initial progress of FIRST. Lattice Logic made available the world's first silicon compiler for random logic [11], and this has evolved into the current product, Chipsmith. They have chosen to support only CMOS (nMOS is less suited to gate-array structures), but designs can be process-independent with that proviso.

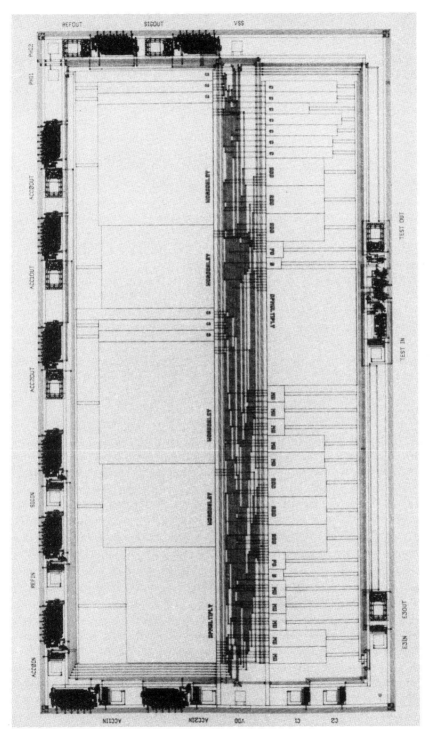

Fig. 11.1   *Floor Plan of an IC produced by FIRST*

Chipsmith lets customers create semi-custom designs on gate arrays, or on 'optimised arrays', or by using 'soft' (parameterised) cell-library techniques. Both gate-array configurations lay out in rows of 'stages', consisting of pairs of p-type and n-type transistors with a common polysilicon (poly) gate.

Use of the standard array allows a customer to independently develop a prediffused gate array image which is particularly suited to his needs, with only the single metal layer to be personalised in the final step. Standard size wiring channels are built from groups of poly underpasses, and surround the 'cores' (columns of stage-rows). Stages are configured with metal interconnect to form the basic library parts, and are wired up with the vertical poly underpasses and horizontal metal.

The optimised array brings improvements in performance, yield etc., at the cost of extra processing requirements. All unused poly or diffusion (including transistors) is removed, and wiring channels are customised, in metal where possible. A similar floorplan is used in the standard cell approach, with standard cells in the rows (instead of stages).

Input to Chipsmith is via the hardware description language Model (Micro-electronic Design Language), from which FIRST was adapted, or by the DraftSmith schematic capture package. Design proceeds in the usual structured fashion of top-down specification and bottom-up implementation. Model supports the decomposition of the design into hierarchical 'parts', down to the level of gates, buffers etc. Some skill is required from the user – he must partition his design with a view to optimal layout, and identify those signals whose timing is critical (these are the first to be routed, and can take optimal paths).

When a Model file has been compiled, it drives both physical design subsystem and logic simulator. The former reads a rule-file for the chosen process, and builds the chip either by connecting between rows of pre-diffused transistors in an underlying automatically-generated gate-array structure, or by custom cell assembly.

Lattice Logic emphasise portability – both of designs between processes and of CAD software between machines. All software is written in IMP (a procedural language peculiar to Lattice Logic), thus reducing the software porting problem to one of writing an IMP compiler for the new machine.

11.6.3 MetaSyn (MetaLogic Inc.)

MetaSyn evolved from a silicon compiler developed at MIT Lincoln Labs, known as MacPitts [12]. MacPitts was targetted at a specific applications area, telecommunications. The user communicates his intent through a LISP-like high-level language interface, and MacPitts composes a custom parallel datapath chip in response. Functional units (adders, incrementers etc.) are interspersed with registers in the datapath, in sufficient abundance to meet the computational requirements of the problem. These units are constructed as a linear array of bit-slice cells known as 'organelles', and the datapath consists of stacked units.

Local communication runs in slots between the slices of organelles and short buses may take advantage of unused slots. As the main datapath is inefficient for boolean (single-bit) operations, these are executed (along with control functions) in a Weinberger array (an array of NORgates). MetaSyn uses a mixture of PLA and Weinberger array techniques for control.

The organelle-library is user-expandable – he must supply the layout and corresponding LISP procedures. Two outputs formats are supported – a technology-independent net-list, and CIF for ARPA/MOSIS nMOS. The main drawbacks of MacPitts were its poor performance in relation to custom CMOS (MetaSyn has improved on this), and the 'arcane' LISP-like user interface. Currently MetaSyn runs on the Symbolics 3600 (a high-performance LISP machine).

11.6.4 Generator development tools (Silicon Design Labs Inc.)
Situated in Liberty Corner, New Jersey, SDL was formed in 1983 by Harold Alles and Misha Buric of Bell Labs, along with key researchers from the PLEX project [13] there. PLEX was unusual in that it was a microcomputer module compiler, taking its input in the form of the actual code which the machine was to execute. PLEX then synthesised an optimal 'core microcomputer', running a minimal subset of instructions compiled from a library.

SDL recently announced their first product, GDT [14], an advanced tool set which allows users to create their own module generators. These generators capture the expertise of their creators, allowing subsequent production of known-good circuits by non-experts. A procedural, object-oriented language L (syntactically related to C) has been developed to capture both structural and geometric aspects of a layout. Three views of the design (language, layout and schematic) are held simultaneously, and the designer may move around these at will. The self-explanatory L-editor and L-simulator are provided to assist in this. All tools are portable and UNIX-based, as one might expect from a Bell Labs spin-off.

GDT is in a sense a 'silicon compiler compiler', and as such might attract interest from IC (rather than systems) houses. This is because any silicon foundry able to offer its customers a suite of generators aimed directly at its own technology could add value to its customers' designs, as well as fabricating them. However the responsibility for correctness then rests on the foundry – arguably the flexibility afforded by the GDT approach leads to loss of the 'correctness by construction' property of turnkey silicon compilers.

GDT would seem to offer powerful design advantages at the relatively low cell or module level. As most of the major workstation suppliers are active in this area, a gradual evolution of the low-level tools (into more powerful, integrated design environments) is foreseen.

11.6.5 Genesil (Silicon Compilers Inc.)

SCI of San Jose CA was formed in 1981 as a spin-off from Intel Corp. (Ed Cheng and Phil Kaufman), along with David Johannsen and Carver Mead from Caltech SSP. The first product from SCI appeared last year in the form of Genesil, a VAX-based turnkey system which automates the entire chip design process. Genesil has evolved from Johannsen's BristleBlocks [6], and has seen several successes to date [15] including the DEC MicroVAX 1 datapath, the Seeq ethernet datalink controller, and the Sun raster-up controller.

Design capture is by forms description – a menu-like input by which the user specifies *and* implements top-down. Default parameters are gradually replaced and refined as the designer homes in on the optimal solution – this is known as incremental design. Genesil supports performance-parameterisation of cells, and has a relatively sophisticated floorplanner.

Genesil can handle the manufacturing parameters of up to twelve foundries, and the design process is technology free until the moment of commitment. Both nMOS and CMOS fabrication are supported. Software runs on the DEC VAX range of minicomputers and workstations, as well as on the Daisy Megalogician.

A recent shift in strategy by SCI resulted in the release of Genesis, a software tool set for silicon compiler creation (similar to GDT). This is a marked departure from the turnkey approach of their earlier products.

11.6.6 Concorde (Seattle silicon technology)

Concorde is a suite of compilers developed by Seattle Silicon Technology (a spin-off from Boeing Inc.) and aimed at the workstation market.

Four main architectural targets are supported: datapaths, random logic, state machines and analog. Like Genesil, Concorde is menu rather than language driven. SST also market a compiler development tool, SLIC.

Initially Valid Inc. had exclusive rights to Concorde – this agreement has now expired and Concorde is also available on other leading workstations.

11.7 The Carver connection

If silicon compilation has a patron saint, it must be Carver Mead of the California Institute of Technology. Of the above named companies, Mead has directly influenced the progress of all but one.

Courses (based on [2]) given by him at Boeing and at Bell Labs resulted in the formation of research groups which spun off into SST and SDL respectively. The Silicon Structures Project which ran at Caltech between 1978 and 1983 involved John Gray of Lattice Logic (as director) and David Johannsen of SCI (as a student). SCI was founded to continue this work, and Mead sits on the board of the company. Other former students of Mead's include Ed Cheng of SCI and Misha Buric of SDL.

11.8 Modern approaches

The silicon compilers mentioned earlier are (with the exception of FIRST) commercially available today. There is also much activity in industrial and academic research labs which will result in the silicon compilers of tomorrow. We identify some of the novel features which these compilers might exhibit.

11.8.1 Behavioural input
To date, silicon compilers which have achieved success have left it to the designer to specify his system as a series of interconnected hierarchical blocks (structural description), rather than what he wishes the hardware to do (behavioural description). BristleBlocks has a low-level structural input, while FIRST is also structural, at higher level. MacPitts (MetaSyn) and the register-transfer language IRENE developed at IMAGE in Grenoble for the CAPRI project [16] are a step towards behavioural input, but the resultant hardware is still architecturally hidebound by the underlying synthesis procedures.

Algorithmic description languages allow the designer to work at an even higher level, simplifying his task but further restricting his freedom. An example of this is CLASP [17], a specificational language for signal processing tasks under development at Bell Labs. CLASP communicates designer intent to SILI, a language-independent silicon compiler which uses data and control flow analysis routines to generate a horizontally microprogrammed datapath machine with the right amount of concurrency for the computational requirements of the task.

Whether or not such advances are to be welcomed depends on whether the translation task (from behaviour to structure) is viewed as one which requires computer automation – that is, whether the task is lengthy, tedious and error-prone.

11.8.2 Floorplanning
Floorplanning to date has been a matter of choosing a 'target architecture', then making the most of available resources within this restricted environment. An alternative approach to floorplanning, based on graph-theoretic techniques, has recently become prominent. 'Idiomatic floorplanning' [18] involves the decomposition of general layout graphs into fragments or 'clusters' which can be optimally planned, and hierarchically assembled to produce the chip. Translation of a planar graph into its rectangular dual [19] may also simplify VLSI floorplanning.

11.8.3 Expert systems
The 'intrusion' of computer scientists into what was once an exclusively electrical engineering activity has resulted in the proliferation of design tools which has culminated in the silicon compiler. Grabbing the bull by the horns in

this manner has allowed computer scientists to have a say in the design of their own machines.

We are about to witness another such intrusion – by the artificial intelligence (AI) community. As special LISP and PROLOG machines begin to emerge, so the application of expert systems to their design points to a next generation of VLSI CAD tools. The 'intelligent designer's assistant' of the future should make the knowledge store of skilled designers available to everybody. Examples of current AI design synthesis research may be found at MIT [20], Carnegie-Mellon University [21] and the University of Illinois [22].

11.8.4 Automatic Feedback
Silicon compilers to date produce layout in an intrinsically open-loop manner. Any feedback loops for layout optimisation must include the designer. The incorporation of feedback mechanisms in the compiler itself [23] should allow the extraction of parametric informaion from the design process, and the use of this information to optimise layout synthesis.

11.8.5 Logic Synthesis
To complete the application span of VLSI design automation, tools for logic synthesis from Boolean expressions are required. Recent activity in the field of PLA logic minimisation (functional and topological) has resulted in efficient tools such as ESPRESSO-II [24]. However the PLA approach yields only two-level logic solutions for combinatorial problems, and fails to take full advantage of 'don't cares'. Multi-level solutions to both combinatorial and sequential problems, aided by AI techniques, are paving the way to the next generation of logic synthesis tools [25].

11.9 Conclusions

The world base of design talent in the systems area is considerably greater than in the more specialised field of IC design. However, until now only a select few skilled IC designers have been able to translate intent into silicon, and even they are beginning to struggle with design complexity. The silicon compiler is proposed as the answer to the complexity problem, automating all low-level IC design tasks, and bringing IC design to within the grasp of systems designers. Only by rapid and correct design synthesis can full advantage be taken of advance in IC technology.

References

1 MOORE, G. E.: 'VLSI: Some Fundamental Challenges', *IEEE Spectrum*, **16**, pp. 30–37 (April 1979).

2   MEAD, C. A., and CONWAY, L. A.: *Introduction to VLSI Systems*, Addison-Wesley (1980).

3   HICKS, P. J. (ed.): *Semi-Custom IC Design and VLSI*, Peter Peregrinus Ltd. for IEE (1983).

4   AYERS, R. F.: *VLSI Silicon Compilation and the Art of Automatic Chip Design*, Prentice-Hall (1983).

5   DENYER, P. B., and RENSHAW, D.: *VLSI Signal Processing – A Bit-Serial Approach*, Addison-Wesley (1985).

6   JOHANNSEN, D. L.: 'Bristle Blocks: A Silicon Compiler', *Proc. 16th DA Conf.* (San Diego, 1979).

7   LEWICKI, G. *et al.*: 'MOSIS: Present and Future', *Proc. MIT Conf. on Adv. Res. in VLSI*, pp. 124–128 (Cambridge, MA, January 1984).

8   BENNETTS, R. G.: *Design of Testable Logic circuits*, Addison-Wesley (1984).

9   MURRAY, A. F., DENYER, P. B., and RENSHAW, D.: 'Self-Testing in Bit-Serial Parts: High Coverage at Low Cost', *Proc. IEEE International Test Conf.*, pp. 260–268 (Philadelphia, October 1983).

10 SMITH, S. G. *et al.*: 'A Comparison of Standard Part and Silicon Compiler Implementations of a Polyphase Network Filter Bank', *Proc. IEEE-IECEJ-ASJ ICASSP'86* (Tokyo, April 1986).

11 GRAY, J. P., BUCHANAN, I., and ROBERTSON, P. S.: 'Designing Gate Arrays Using a Silicon Compiler', *Proc. 19th DA Conf.*, pp. 377–383 (Las Vegas, 1982).

12 SOUTHARD, J. R.: 'MacPitts: An Approach to Silicon Compilation', *Computer*, **16**, pp. 74–82 (December 1983).

13 BURIC, M. R., CHRISTENSEN, C., and MATHESON, T. G.: 'The PLEX Project: VLSI Layouts of Microcomputers Generated by a Computer Program', *IEEE Intl. Conf. on Computer-Aided Design* (Santa Clara, CA, September 1983).

14 BURIC, M. R., and MATHESON, T. G.: 'Silicon Compilation Environments', *Proc. IEEE CICC*, pp. 208–212 (Portland, OR, May 1985).

15 JOHNSON, S. C.: 'VLSI Circuit Design Reaches the Level of Architectural Description', *Electronics*, pp. 121–128 (3rd May 1984).

16 ANCEAU, F., and SCHOELLKOPF, J. P.: 'CAPRI: A Silicon Compiler for VLSI Circuits Specified by Algorithms', pp. 149–154 in *VLSI Architecture*, ed. B. Randell & P. C. Treleaven, Prentice-Hall (1983).

17 KAHRS, M.: 'Silicon Compilation of a Very High Level Signal Processing Specification Language', pp. 228–238 in *VLSI Signal Processing*, ed. P. R. Cappello *et al.*, IEEE Press (1984).

18 DEAS, A. R., and NIXON, I. M.: 'Chromatic Idioms for Automated VLSI Floorplanning', *Proc. VLSI '85*, pp. 55–64 (Tokyo, August 1985).

19 KOZMINSKI, K., and KINNEN, E.: 'An Algorithm for Finding a Rectangular Dual of a Planar Graph for Use in Area Planning for VLSI Integrated Circuits', *Proc. 21st DA Conf.*, pp. 655–656 (Albuquerque NM, June 1984).

20 SHROBE, H. E.: 'AI Meets CAD', *Proc. VLSI'83*, pp. 387–399 (Trondheim, August 1983).

21 THOMAS, D. E. *et al.*: 'Automatic Data Path Synthesis', *Computer*, **16**, pp. 59–70 (December 1983).

22 HEALEY, S. T., and GAJSKI, D. D.: 'Decomposition of Logic Networks into Silicon', *Proc. 22nd DA Conf.*, pp. 162–168 (Las Vegas, June 1985).

23 FRIEDMAN, E. G.: 'Feedback in Silicon Compilers', *IEEE Circuits & Devices*, **3**, pp. 15–20 (May 1985).

24 BRAYTON, R. K. *et al.*: *Logic Minimization Algorithms for VLSI Synthesis*, Kluwer Academic Publishers (1984).

25 NEWTON, A. R.: 'Techniques for Logic Synthesis', *Proc. VLSI '85*, pp. 27–39 (Tokyo, August 1985).

Chapter 12

CAD systems

J. D. Wilcock

12.1 Systems

Electronic systems are composed of a variety of parts, mechanical, electrical and electronic. Such parts include boxes, castings, aerials, frames, consols, printed wiring boards, hybrid circuits, backplanes, commercially available integrated circuits, custom integrated circuits and discrete components. In addition there is software and firmware. The complete systems also has to meet many design requirements, such as power, weight, performance, temperature, radiation and vibration resistance.

Computer aids can be used for the design, test and manufacture of each of these parts. The completely computerised design and manufacturing system of the future would integrate the various CAD/CAM/CAT systems for the parts, together with other computer programs used by the manufacturing process, such as stock control and finance.

It is clear that the CAD systems for IC design are a small part of the total CAD systems required.

Also, it can be observed that the design activities for all these systems has something in common. It is useful for the designer of IC CAD systems to recognise the common elements in all design activities.

12.2 Factors common to all design activity

Four factors can be recognised that are part of every design activity. These are

(1) Starting specification capture. This involves taking a specification of the part to be designed, and entering it into the design system. The specification could come from a customer, or from a higher level part of the design system.

(2) Design. The design activity itself involves adding information to the specification. There is an extremely large number of possible outcomes of this activity. In practice the field of search for a particular outcome is restricted by design constraints. The constraints may be of two kinds, codes of practice and rules, and the sharing of resources. In the case of IC design, the rules may include timing rules, or layout rules. The shared resource may be chip area or power available.

(3) Verification. The completed design must be checked against the starting specification. In verification there are only two possible outcomes, that the IC meets the specification, or does not meet the specification. In practice, if the circuit does not meet the specification, then the design engineer needs diagnostic information, to enable him to modify the design. The verification should ideally identify all the defective parts of the design, without identifying as suspect any of the correct parts of the design.

(4) Output. The output may be to manufacture, or to another lower level of the design process.

These four activities are necessary in every design process, including manual design processes. In the design of an IC CAD system it is important to recognise which software is being used for which of these four activities.

Fig. 12.1 shows the hierarchy of these design activities.

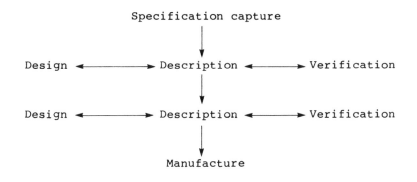

Fig. 12.1   *Hierarchy of design activities*

It is clear that design and verification are closely related. Once CAD systems are at such a state of development that both design and verification are done on the same computer, it is possible to arrange for on-line checking of the design as it proceeds. This checking task can be carried out by the computer during the time that the designer pauses for thought. This gives the designer an immediate response on the correctness of his design. It has a particular advantage when it comes to the diagnosis of the cause of any errors, as the last entry made by the designer is often the cause of the error.

The ultimate verification capability is provided by the long term objective of formal design methods. The approach is that the specification is described in a mathematically correct way, perhaps in some form of algebra or calculus, and the design is derived by mathematical proof techniques. As long as each step is proved, there is no need for additional verification at all.

12.3 Evolution of IC design systems

In the earliest IC design approaches circuit design was a manual activity. The specification was a paper document, and design was by thought and mathematical analysis using pencil and paper. The design was verified by building a breadboard model, and using oscilloscopes and other electronics instrumentation to show that the response of the circuit to stimuli was that desired.

For layout the circuit design was effectively recoded, using coloured pencils and square paper. The circuit was recorded again into cut and strip Rubylith, ready for photographic reduction into the form required for mask making. Verification of the Rubylith was by inspection by eye by experienced layout engineers. This checked that the Rubylith represented the original drawing. Verification of the photographic stages was by examination of the plates and masks by eye under a microscope. This checked that the mask represented the Rubylith.

As IC CAD systems developed, computers were used for controlling the mask making pattern generators. These were fed by graphics computers, such as Calma and Applicon, which included facilities for capturing layout data, by manual input, and for checking the data against geometric layout rules. Computers were also used at the start of the design process for analytical predictions of circuit performance.

The result was that many IC CAD systems still are a mixture of computer programs, involving manual recording of data at some point in the design cycle. As manual coding is involved, it must always be checked by verification programs, which, for VLSI circuits, can consume considerable computer resources. The use of mixed computer and manual IC design leads to many problems, such as control of libraries, and proliferation of software bridging programs. Libraries of information are required for all design and verification tools. For IC CAD the tools include process, device, circuit, logic and functional simulation, layout tools and verification tools.

New software that is added to the CAD system must be interfaced to several programs in the existing system. This leads to many bridging programs being produced, and many interchange formats.

12.4 Current requirements for IC design systems

The first requirement for a modern IC design system is that it is integrated. That is to say that, once the data is entered into the system, the same information does not need to be entered again.

Secondly the same data may need to be obtained from the CAD system in several forms. Different CAD programs may need different formats of data. The different classes of user (Chip Designer, Cell Library Designer, Silicon Vendor) may require the same data in different formats.

The data describing the chip design needs to be consistent. If the same information is held in two places in a CAD system, there is always a chance that one copy may be altered without alteration of the other. This can lead to incorrect IC design, due, for example, to running verification on a different circuit description to that used for manufacture.

The CAD system should not force a strict design route on the designer. Different designers might want to carry out their design activities in different order. The IC CAD system should not force a particular sequence on the designer.

The CAD system should be capable of top down as well as bottom up design. In top down design the designer partitions his resources (such as chip area) among the parts of the chip yet to be designed. In bottom up design (or implementation), the components forming the chip have all been designed, and have to be assembled together (by placement and routing) to form the chip.

The ability is needed to mix bought-in software with in-house software. This requires that new software tools can be easily added to the CAD system.

The system must be able to handle sudden specification and design changes which often occur during the course of the design. This implies that parts of the circuit unaffected by the change need not be updated when the change is implemented. The ability to handle changes becomes more important as the complexity of VLSI circuits increases, for the computer and human resources for a given stage in the design increase rapidly.

Sometimes the system designer requires a change in an IC design several years after the original design was completed. By this time, different CAD tools are probably in use. The IC design system must be able to restore the CAD tools and libraries as they originally were, and then translate the design to the current CAD tools, to avoid having to redesign the circuit from scratch.

12.5 The data base management system as a solution

For the IC CAD system these objections can be met by having a central repository for one copy of the design data. This meets the requirements for recording and consistency. To avoid forcing a strict design route on the users, the data repository should have slots for adding in the new data created as the design proceeds and there should be no need to fill these slots in any particular order. There is commercially available software which meets these requirements, and that is the Data Base Management System. The features that are useful for general Data Base Management System applications are also useful for IC CAD applications.

12.6 Features of data base management systems

A definition of a database is that it is a collection of interrelated data stored together, with controlled redundancy, to serve one or more applications, the data are stored so that they are independent of the programs which use the data; a common and controlled approach is used in adding new data and in modifying and retrieving existing data within a database.

The data definition is centralised in a Data Base Management System, rather than being distributed among various application programs. This ensures that all applications programs have access to any data they require. It is easy to interface applications programs to a Data Base Management System. The database provide facilities to allow the application program to view the data in a form that is suitable to the program. Writing of parsers and lexical analysis programs, necessary when language interfaces are used between CAD applications programs, is avoided. The number of interface programs to be written is also reduced. If there are N applications programs, required to be interfaced to each other then without a central data format $N*(N-1)$ interface programs must be written. With a central data format, only $2*N$ programs are needed.

With centralised data it is important to provide protection against data corruption. Data Base Management Systems normally provide protection against hardware failure, software failure and unauthorised access. Protection is normally provided, even if a hardware fault occurs in the middle of a long transaction between an applications program and the Data Base Management System, for example, when the Data Base Management System has only been part updated. Concurrent use of the data is also normally catered for, by locking mechanisms. Locking mechanisms prevent one user accessing data which may be incorrect or inconsistent due to its being changed by another user.

Different parts of the data may be protected against access or update by particular users. This may be required for libraries, or for other commercially confident parts of the CAD system.

The centralised nature of the Data Base Management System allows the useage of libraries to be controlled, so that the correct issues are used in particular designs. Another advantage of centralisation is that it allows better visibility of the progress of the design work both for the designer and for the manager.

Data Base Management Systems often allow mixing of data from different databases. This is important for networked IC design systems, where an application program on one node of the net may require data from several other nodes.

12.7 Differences between IC and general data base management systems

The requirements for a Data Base Management System for an IC CAD system do differ from those of a commercial Data Base Management System. In an IC Data Base Management System an application program normally access and updates a large volume of data, in fact most of the data in the database. In a commercial application of Data Base Management System only a very small part is updated in a transaction. In a CAD system, the application programs take a long time to run, perhaps hours or longer. In a commercial Data Base Management System the transactions are much shorter. In a CAD Data Base Management System the queries to the database are rarely updated. In a commercial application, queries may be changed quite frequently, and a good query languaged is needed.

There are several types of Data Base Management System available. For IC CAD work, the Relational Data Base Management System is most suitable. In comparison with other types of Data Base Management System, the relational Data Base Management System is easier to set up, and easier to change and update the data model. The relationships in the data model are not built into the access mechanisms, and so can be changed easily. Other types of Data Base Management System offer faster access, but changing the data model is more difficult.

12.8 The nature of IC Data

Although CAD tools can become out of date rapidly, the fundamental data objects they handle change much more slowly. Therefore to design a CAD system for long life, it is often advantageous to start by considering the fundamental nature of the data, rather than by considering particular tools. This section considers the nature of IC data.

There are three classes of IC design data involved in describing a VLSI integrated circuit, namely behavioural, structural and physical.

Behavioural data is what the circuit does. It concerns the response of the circuit to stimuli. Programming languages and test pattern truth tables are examples of behavioural data.

Structural data describes what the circuit is made of, in terms of the electrical interconnection of cells or other building blocks. The structural description can be hierarchical, when each cell contains further structure within it. Network coding languages for simulators and autorouters are examples of structural data.

Physical data describes such things as positions and orientations. The positions and orientations may be of symbols or polygons. Symbols are physical abstractions such as gate symbols on a logic schematic, cell outlines in a cellular design system or the sticks in a sticks layout tool. Physical data makes up the final output of IC layout CAD systems.

Fig. 12.2 illustrates these classes of data. The IC design process generally starts with a behavioural description of the desired circuit, and certainly ends with a physical description of the circuit. However, in the course of the design process all representations of the circuit can be present simultaneously.

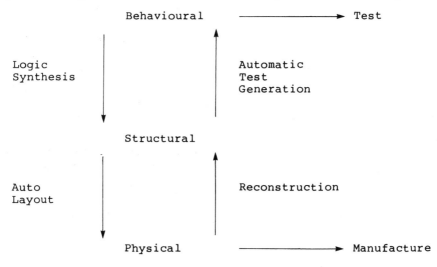

Fig. 12.2   *Classes of data for integrated circuits*

CAD design tools can be considered as mapping between the various representations in a given class of data, or as mapping between classes of data. Some examples of mapping between classes are given below.

Automatic mapping from behavioural description to structural description is by means of silicon compiler software. The Paracell Assembly software used in the PLESSEY MEGACELL (TM) system is an example. For a PLA the input is a behaviour, expressed in terms of Boolean equations or a truth table. The output is a structural description consisting of an interconnection of component cells.

Automatic mapping from structural to physical descriptions is via autoplacement and autorouting software.

Mapping from physical back to a structural description is via layout reconstruction software, which produces a switch level or logic level description from a physical layout.

Mapping from structural to behavioural descriptions is done by conventional automatic test pattern generation software. The input is the structure. From this the fault list is derived. The output is a truth table, covering as many faults as possible.

12.9 Preparing a Data model

To make use of a Data Base Management System, it needs to contain a description of the data being used. This is called the data model. A data model is an abstract view used to define the inherent characteristics of the data items and their relationships in a precise and consistent fashion. One data model can have many physical representations.

If data modelling is done well it minimises future disruption in response to changes in applications programs, such as CAD tools, and in response to changes in Data Base Management System Software, and the hardware.

The data model we require is for a Relational Data Base Management System. However, if a data model is in a suitable form for a Relational Data Base Management System, it will also be suitable for other types of Data Base Management System.

A Data Base Managment System allows each application program to have its own view of data (subschema). This need not be the same as the Data Base Management Systems view (schema). The Data Base Management Systems view should have maximum simplicity and consistency, with no duplication of data. Our task is to derive the Data Base Management System view of the data.

We are talking about a logical view of the data, not a physical view. As far as possible, the physical implementation should be disregarded. Ideally, the Data Base Management System makes the logical and physical views of the data independent. The applications programs have one logical view. The Data Base Management System has another logical view. And the machine operating system handles the physical representation.

We should expect to be able to change the data model, while leaving the applications programs the same, because the logical view is independent of the logical view of the stored data (logical data independence).

We should also expect to be able to change the physical layout and organisation of the data, without changing either the overall logical structure of the data or of the applications programs.

The remainder of this section describes approaches for analysing the data, to obtain the optimum data model. Further information on Data Analysis and Data Modelling can be found in the References.

The first point to note is that CAD application programs change, and are updated at frequent intervals. The data objects that are used in CAD programs change much more infrequently. Therefore it is advisable to derive the data model by consideration of the data, rather than by consideration of a particular set of applications programs.

Data objects can be divided into two kinds, entities and attributes. An entity is an object about which we wish to store information. An attribute is some information about an entity. Eg an entity could be 'employee', and an attribute could be 'salary.' It is important to distinguish between entity and attribute type and instance. The above examples are types. Examples of instances would be 'Joe Bloggs' and '$15000' respectively.

The result of the data analysis should be a model in what is called 'third normal form'. The definition of normalisation arose from the work of E. F. Codd. The first normal form means arranged as a two dimensional table, as are records in a conventional (physical) computer file. The second and third normal forms are rearrangements of the table into simpler and more consistent tables.

James Martin summarises his chapter on third normal form, with the definition 'All data in the record are functionally dependent on the key of that record, the whole key and nothing but the key.'

Table 15.1 shows some IC CAD data objects arranged in a table in first normal form. The cell instance is a particular example of a cell type. Each cell type comes from a library, and has an issue number. There are several problems with data arranged in this form. The information about the cell type, such as the size, is repeated many times. This would increase the time to update the database. Also, you cannot hold information about the cell size unless it is placed. Deleting the last instance of a cell during an editing session would remove information about the cell.

One approach to analysis of this data is to find which attribute is functionally dependent on which. The cell type is dependent on the cell instance, because for any cell instance there can only be one cell type. Similarly, for each cell

Table 12.1 *Data in first normal form*

Cell Instance	Cell Type	Issue	Size	Orien- tation	Position
X123	NOR	1	50*50	0	1000,1
X124	AND	1	50*50	1	200,1
X999	NOR	1	50*50	2	300,1
X765	DTYPE	2	250*50	0	600,99

type there can only be one cell size. The dependencies are marked in Fig. 12.3a. From this figure it can be seen that the size depends on the cell type which depends on the cell instance. Therefore the dependency of size on instance is redundant, and can be removed (Fig. 12.3b). It is now seen that some attributes depend only on the cell type, and some depend only on the cell instance. The table should therefore be split into two tables, as shown in Table 12.2. This has overcome the problems with the original table.

Another technique which also results in a data model is to produce a canonical schema. The canonical schema is a minimised data model, which is derived from the views of the data seen by the various applications program interfacing to the database. This is done by building up a global data model.

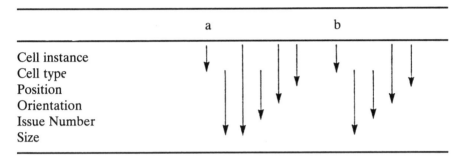

Fig. 12.3 *Functional dependence diagram*

Table 12.2 *Data in third normal form*

Cell Instance	Cell	Orien- tation	Position
X123	NOR	0	100,1
X124	AND	1	200,1
X999	NOR	2	300,1
X765	DTYPE	0	600,99

Cell Type	Issue	Size
NOR	1	50*50
AND	1	50*50
DTYPE	2	250*50

The data model for each application program in turn is examined, and put into third normal form. It is then added to the global model, which is then simplified, if possible, by, for instance, eliminating redundancies. The definition given by Martin is 'Canonical schema: a model of data which represents the inherent structure of that data and hence is independent of individual applications of the data, and also of the software or hardware mechanisms which are employed in representing and using the data'. A canonical schema will always be in third normal form.

During the analysis of the data, it is useful to keep checklists. These lists were originally suggested by ICL.

Attribute checklist: Definition: name, description, synonyms, data type and characteristics; Where used: (different entities); Statistics: Population and growth rate; Authority: who put the data in; External formats: for source documents and reports; Privacy: who can access? Range of values; Notes:

Entity checklist: Definition: name, description and attributes (including entity identifier); Connections: attribute which relate this entity to other entities; Statistics: No. of occurrences, growth rate; Authority: who responsible for; Where used: Privacy: Archival: number of versions and length of time an occurrence is required to exist; Dependent attributes: intra-entity structure, ie dependency of one attribute on another; Notes.

The checklists can be formalised in a data dictionary. This is particularly important when there are several users, who might, for instance, have synonyms for the same data object. The data dictionary can also record information on which data objects are used by which applications programs, so that when the data model is updated it is known which applications programs must also be updated.

Finally, an aid to discovering all the attributes associated with a particular entity is to draw an attribute/operation matrix. Table 12.3 shows such a matrix. Each row is an attribute. Each column is an operation. Each position contains C U R D or blank (created, updated, read, deleted or blank).

The example shown contains part of the attribute/operation table for an IC CAD database. It can be seen that, of the data objects mentioned, simulation only requires knowledge of the cell issue number. Placement requires to read

Table 12.3 *Attribute/operation matrix*

	Simu-lation	Place-ment	Route	Update library
Cell Size		R	?	U
Cell Issue No	R	R	R	U
Cell position		C	R	D
Track layer			C	?

the cell issue number, and the cell size; it creates the cell position. Routing requires the cell position, and creates the interconnection track layer. However, whether it requires knowledge of the cell size or not cannot be answered, unless the relationship between pins and cells is clearly defined elsewhere.

Similarly, if a cell library is replaced by an updated cell library, the following should happen. Previous cell placements should be deleted. The cell issue number should be updated. The table has highlighted that something must happen to the track layer, but what exactly is an implementation decision.

12.10 Programs interfacing to the IC data base

There will be many applications programs interfacing to the IC CAD Data Base Management System. Some of these are listed below.

Network language data entry
Schematic entry
High level simulators
Autolayout tools
Floorplanner
Interactive layout tools
Checking tools
Simulators
Silicon Compilers
Output, plotters, log files
Output to manufacture
Interface to other data bases (EDIF etc)

12.11   Conclusions

Similarities are observed between many different types of design system. By learning from such experience, and from the experience of the evolution of IC design styles, it is concluded that a database management system should be at the heart of a modern IC CAD design system. A database management system cannot be used without a data model. The task of preparing such a data model is described. The outline of a typical CAD system is given to show the types of programs interfacing to the database.

References

1 MARTIN, JAMES: 'Computer Data Base Organisation, 2nd edition' 1977, Prentice Hall
2 DEC VAX Rdb/VMS documents (Guide to Database Design and definition, Reference Manual, Summary description, Common data Dictionary)

Chapter 13

Data management

D. Warburton

13.1 Introduction

Traditionally, the production of a new CAD system has seen design and implementation effort concentrated on the provision of application tools; data and its control often being relegated to a poor second place.

However, the challenge facing today's electronics industry is the competitive exploitation of hardware technologies which will permit chip densities of 250K transistors now and >1 million transistors in the near future. The emphasis for such VLSI chips will be 'speed into the marketplace' with the obvious implication of 'get it right first time'.

The CAD system will be required to support hierarchic design styles, greater emphasis on design verification, new approaches to testing and so on. However, all these enhancements will be pointless, if time, effort and money are wasted, enhancing, verifying, or even manufacturing the wrong or inconsistent versions of the design. In future, the management and control of the design data will be of equal importance to the CAD tools themselves.

13.2 Requirements

If the management and control of data is to be implemented successfully then the requirements to be met by the data management system must be clearly defined. The following are seen as essential for an ideal CAD data management system:

★ Provide the design data for all application tools via a set of common interfaces.
★ Control all access to the design data.
★ Support a common user interface to all data.

★ Efficient. Interactive applications can be interfaced directly to the DMS.
★ Incorporate the required mechanisms for design state control in the DMS – not build-on afterwards.
★ Provide accounting, privacy and security.
★ Resilient. Data must revert a known state following a machine break. Automatic and controlled archiving to allow rebuilding of completed data.
★ Retention of data integrity across external interfaces. These interfaces may be to alien CAD systems, silicon vendors etc.

A forward-looking data management system should also meet the following criteria:

★ Independent of host system. The DMS and the data contained may be easily transferred between different machines.
★ Capable of operation on a networked system. Transparent distribution of data over a network of similar or differing computing nodes.

13.3 Alternative approaches

13.3.1 File-based systems
Virtually all current CAD systems, either commercially-available or in-house, fall into this category. Data is stored in serially-accessed files, many of which are application specific. Each application provides its own routines to read interpret and create the data held in these files. Communication between applications is via one or many such files, either directly or via various stages of inversion.

Advantages:

1. Quick and easy to implement.
2. Data accessed by edit/interactive tools can be tailored for performance.
3. Can make good use of host facilities.
4. Inherent resilience.

Disadvantages:

1. Much data duplication.
2. Inconsistent interpretation of data.
3. Difficult to enhance data descriptions – prone to structural decay.
4. Not amenable to hierarchic design structures.
5. Control difficult.
6. Dependence on host facilities.

ICL's DA4 system (Adshead [1]) is an example of a file-based system. It was originally designed in the mid-seventies to handle hardware designs built using regular PCBs containing well-defined MSI and SSI components where a flat design style was adequate.

It was subsequently successfully enhanced to process designs based on gate-array technologies using limited hierarchic design style. Again the basic cell design components were well-defined and fixed.

13.3.2 Integrated database systems

Here all the design data is held within and controlled by the data management system (see Fig. 13.1). All applications interface directly with the DMS.

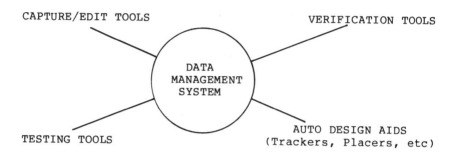

CAPTURE/EDIT TOOLS   VERIFICATION TOOLS

DATA
MANAGEMENT
SYSTEM

TESTING TOOLS   AUTO DESIGN AIDS
(Trackers, Placers, etc)

Fig. 13.1   *An integrated database system*

Functionally, most commercially-available network database systems will meet the DMS requirements for CAD. However, experience in general has shown their performance, when handling CAD data, to be poor. Also, use of a commercial database would normally make the CAD system host-dependent.

In practice, then, we can expect the integrated DMS system to be constructed around a purpose-built infrastructure.

Advantages:

1. No data duplication necessary (except for controlled archive purposes).
2. Common data interface is possible and desirable.
3. Easy representation of hierarchic structures.
4. All relationships between design components (and design levels) can be recorded explicitly.
5. Offers maximum control capability.
6. Potential for host independence.

Disadvantages:

1. DMS infrastructure must be in place first.
2. Need better control over development and release of CAD software.
3. Less scope for tailoring for performance (but its ability to rush headlong towards disaster an advantage?).
4. If host independence desirable, then may need to duplicate host facilities.
5. Resilience must be designed in.

ICL's DA-X system (Harwood [2]) is an example of an integrated database CAD system. Work on DA-X commenced in 1983 with an initial emphasis on standard cell-based VLSI designs and free-format PCBs. Following the successful completion of test chips designed using DA-X, the system is now in use, in ICL, for the processing of real VLSI chip designs.

13.4 Design control in DA-X

In order to manage the complexity of a large VLSI chip, ICL has adopted a hierarchic design style. DA-X supports a true multi-instance hierarchical design breakdown.

Any node within the hierarchy may be considered as a unit of design in its own right (see Fig. 13.2).

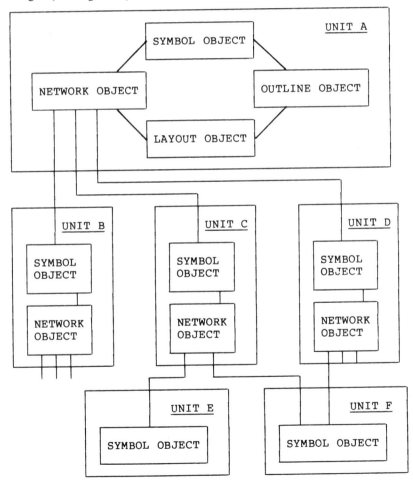

Fig. 13.2 *Simplified DA-X hierarchy*

A unit of design may be built from other units of design which, in turn, may be built from still further units and so on.

Each unit contains a set of views. The four principal views are:

Symbol – this defines the unit's interface to the design hierarchy.

Network – this is the schematic diagram for the unit which, in turn, instances symbols of lower level units.

Outline – this defines the unit's interface to the physical hierarchy supplying its physical shape, terminal positions, etc.

Layout – this is the floorplan for the unit and instances outlines of lower level units.

In DA-X, views of a unit are held in objects. The symbol, network, outline and layout objects are stored in a structured format to enable the inter- and intra-hierarchy relationships to be explicitly recorded. Other views of a unit (e.g.: Test Method, Behavioural Model) can also exist (in either structured or text format) and will be referenced to and from the appropriate principal object. To differentiate between design iterations on an object, each existence of an object is allocated a unique (to the system) state.

1.3.4.1 Development phase

During the development phase DA-X permits the different views of a unit to be developed independently of each other and hence, tolerates incomplete and/or inconsistent design. Similarly, inconsistencies must be tolerated between levels in the design hierarchy.

However, the final objective is to produce a totally consistent design both across and within all levels in the hierarchy. Therefore, it is essential to give the designer facilities to determine what inconsistencies exist in his design and where they are.

The error information is achieved by running the appropriate DA-X error checking software. This will associate the errors detected with the specific states of the relevant objects. Hence, associated with every state of each object will be a set of errors.

Consider first the symbol object. The symbol is defined as the specification for the unit and is therefore 'master'. Any inconsistencies between itself and, say, the network object for the unit will be recorded with the network. Any errors recorded with the symbol will only relate to the symbol itself.

The error status of the symbol is therefore limited to one of the following:

Not checked.
Checked: No Errors.
Checked: Errors Found.

This error status will always remain valid for a specific state of the symbol.

Consider now the network object. The network references other objects within the set of views (objects) defining the unit of design. In addition, it instances the symbol objects of lower level units in the hierarchy. A change to any of these referenced objects can affect the network and yet the state of the network will not have changed. Further, the error reports produced by previously run checks and associated with the network, may no longer be valid.

The error status of the network may therefore be any of the following:

Not Checked.
Checked: No Errors/Valid.
Checked: No Errors/Invalid.
Checked: Errors Detected/Valid.
Checked: Errors Detected/Invalid.

To control this situation error checking software will associate errors detected, not only with the state of the object being checked, but also with the states of all referenced objects.

As these two examples show, the DA-X DMS provides an inbuilt mechanism for controlling design data during development. With the aid of interactive query commands in DA-X, the hardware designer can determine the validity of his design at all stages.

13.4.2 At release time

As we have seen, during development, the DA-X objects which contain the different views of a unit of design have the freedom to change independently of each other. Similarly, objects contained in the lower level units from which the unit is built are also free to change.

Obviously, such fluidity cannot be permitted in a design which is to be released for manufacture. No unit in the hierarchy can be a candidate for release until the designs of the underlying units from which it is built have first been released and frozen. Therefore, in a hierarchic design, release must be a progressive bottom-up process. Following release, not only will the set of objects which define the unit be frozen but also the inter-object relationships will be fixed to show the frozen states of the objects referenced.

However, before a unit can be considered for release it must be demonstrated to be consistent within itself and with the underlying units from which it is built. The same mechanism in the DA-X DMS which enabled the validity of a design to be controlled during development will be used to provide reports for a design audit. These reports will indicate what has been checked, the results obtained and whether they are still valid. Any residual valid errors may be waived by the appropriate level of design authority (electronic sign-off).

As part of the release process, all objects containing the design views of the unit will be moved from the development area in the DA-X DMS to a secure release area. Deletion locks will be applied to prevent their accidental erasure. All manufacturing data will be derived from this secure data.

13.5 Conclusion

The infrastructure of the DA-X data management system provides a suitable foundation for the achievement of our original list of requirements for the ideal system.

It has been demonstrated that the comprehensive control mechanisms required during the development stages of a VLSI design can be readily supported. In future, the inherent control capability of the integrated DMS will be extended to cover such areas as: external interfaces; better integration of alien CAD software or systems, and management of the manufacturing interface. The latter is of particular importance for PCBs, where the control of modifications must be maintained following manufacture.

The successful development of products using the evolving hardware technologies will place great demands on CAD systems in the future. Control will be of paramount importance as design complexities continue to increase. DA-X has recognised this importance and adopted the integrated solution from the outset.

References

1. ADSHEAD, H. G.: IEE, EDA, 1981, 1–4
2. HARWOOD, D. C.: IEE, EDA, 1983

Chapter 14

Verification of digital systems

M. H. Gill

14.1 Introduction

In the field of Software Engineering the practical application of Formal Verification is still a triumph of hope over experience. At the outset, it has therefore to be recognised that the application of these same Formal Methods to digital system design is still in its infancy. Nevertheless, significant progress is being made and it is the purpose of this chapter to review three relevant approaches in order to highlight that progress.

The area of Formal methods is rooted in mathematics and is therefore amenable to a very mathematical presentation. However, a more pragmatic presentation will be employed here. This is not intended to undervalue the importance of theoretical base but to open the subject to the less specialist reader and to give a feel for the current state of the art.

14.2 What is verification?

The unthinking meaning which is frequently ascribed to 'Verification' is, in the hardware context, 'To ensure that the circuit is correct', A moment's thought reveals that this, at the very least, requires the ability to read the mind of the original specifier just to determine what the circuit should do.

A more realistic concept of verification requires a clearer notion of 'Rightness'. The notion of rightness used here depends on the 'contractual model of design', Denvir (1). The design process is conceived as a series of 'Contracts'. Each contract is an agreement by a contractor to develop an implementation which conforms to a specification presented by the specifier.

The rightness of the original specification cannot be proved. It may be possible to deduce that the specification has or has not undesirable properties such as deadlock. It may even be possible to show that the specification is inconsistent. It may be possible to animate the specification in some way, but the most that this can achieve is that the specifier confirms the behaviour demonstrated under the conditions applied.

The implementation is 'right' if it conforms to the specification. Conformance in this sense can be in terms of strict equivalence; that is, the implementation can be shown to have the same behaviour as the specification in some algebra and the two are therefore semantically equivalent even though they differ syntactically. Alternatively, the implementation may be shown to be equivalent to the specification under some limited conditions.

The process of verifying the implementation can be achieved by determining an implementation and then attempting to show conformance with the specification. However, even with concise descriptions the volume of the detail involved is usually daunting.

One step towards containing this problem is to extend the contractual model and apply it to successive levels in the hierarchy of the design description. This orders the verification process as a series of small steps. This is of itself insufficient and there is a need for support tools which will implement at least the elementary steps involved in verification.

In order to show conformance these tools have to operate on both specification and implementation. Traditionally these tools have been conceived as theorem proving tools. Recent work in software verification has developed the idea of transformation tools. That is, tools which either implement transformations that preserve the semantic content of the description while altering its syntactic structure or which implement semantic altering transformations in a controlled manner.

14.3 LSM

LSM, Logic of Sequential Machines, is an early example of a system designed to show that the verification of real hardware is practical. Real examples require detailed descriptions and the construction of proofs requires machine support if it is not to become a process which induces errors by its sheer tedium. This support is provided to LSM by embedding it in a programming environment called LCF. This brief description will not include details of LCF. A more substantial treatment of LSM can be found in Gordon (2).

Behaviour expressions in LSM have the form

NAME (state variables) =
  λ\{input line Names\}.
    \{Output line names = Expressions giving outputs\},
      NAME (Expression giving state at next time step)

To illustrate the meaning of this consider a latch circuit. Informally, this can be described as a circuit whose inputs are WRITE and INPUT and whose output maintains its value until the write signal is true at which time the output takes the value of the input.

In LSM this would be expressed:

$$\text{LATCH(S)} = \lambda\{\text{Write, Input}\}.$$
$$\{\text{Out} = (\text{Write} \rightarrow \text{Input, S})\},$$
$$\text{LATCH (Write} \rightarrow \text{Input, S})$$

The construction 'Write → Input, S'
is to be read as 'if Write is true then Input else S'.

Thus the output of the device is either the stored state or the input signal if the write line is true. Similarly the state will be changed at a particular time step if the write line is true.

If now we consider a slightly more complex latch, one which has both read and write signals, then its behaviour can be defined by the following:

$$\text{REG(SW, SR)} = \lambda\{\text{R, W, I}\}.$$
$$\{\text{Out} = (\text{R} \rightarrow \text{SW, SR})\},$$
$$\text{REG ((W} \rightarrow \text{I, SW), (R} \rightarrow \text{SW, SR))}$$

essentially the state has become more complex as the output value may not be the signal captured by the most recent write signal.

This more complex latch could be composed from two simpler latches, as in Fig. 14.1.

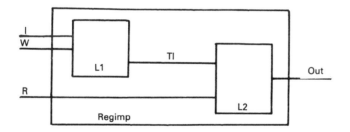

Fig. 14.1   *A constructed latch*

Individually the behaviour of the two simple latches can be described as:

$$\text{L2(S2)} = \lambda\{\text{R, TI}\}.$$
$$\{\text{Out} = (\text{R} \rightarrow \text{TI, S2})\}, \text{L2 (R} \rightarrow \text{TI, S2})$$
$$\text{L1(S1)} = \lambda\{\text{W, I}\}.$$
$$\{\text{TI} = (\text{W} \rightarrow \text{I, S1})\}, \text{L1 (W} \rightarrow \text{I, S1})$$

In order to join these devices together to get the device Regimp of Fig. 14.1 we need to:

(a) Connect all outputs and inputs with the same name.
(b) Hide all the internal signal lines.

Using the composition theorem of LSM the two components compose to give the following behaviour description where TI is the only signal to be hidden.

$$\text{Regimp (S1, S2)} = \lambda\{R, W, I\}.$$
$$\text{Letrec } \{TI = (W \rightarrow I, S1),$$
$$\text{Out} = (R \rightarrow TI, S2)\}$$
$$\text{in } (\{Out = (R \rightarrow TI, S2)\},$$
$$\text{Regimp } ((R \rightarrow TI, S2), (W \rightarrow I, S1)))$$

Unfolding this gives:

$$\text{Regimp (S1, S2)} = \lambda\{R, W, I\}.$$
$$\{Out = (R \rightarrow (W \rightarrow I, S1), S2)\},$$
$$\text{Regimp } ((W \rightarrow I, S1), (R \rightarrow (W \rightarrow I, S1), S2))$$

This simplifies to:

$$\text{Regimp (S1, S2)} = \lambda\{R, W, I\},$$
$$\{Out = (R \rightarrow S1, S2)\},$$
$$\text{Regimp } ((W \rightarrow I, S1), (R \rightarrow S1, S2))$$

So it can be seen that Regimp and REG have the same behaviour because the states S1 and SW are equivalent, as are the states S2 and SR.

This example represents, in sketch form, the way in which LSM can be used to deduce the behaviour of a constructed item from the detailed behaviour of its parts. The application of this to larger, hierarchically structured, designs is obvious. Gordon (3) demonstrates the use of LSM to verify a small microprocessor. Gordon (4) applies the ideas to several smaller modules. Barrow (5) describes a PROLOG based implementation of these ideas which has successfully verified large designs.

The LSM system is limited in that the theorems available to the user are built in to the software and their formal function is not clearly visible. This has led to the development of the more formally based, more extensible system; HOL.

14.4 HOL

HOL represents a change from the special purpose logic of LSM to a more general Higher Order Logic, hence the name; HOL.

Higher Order Logic is perhaps most simply described as a generalisation of first Order Logic. In first order logics variables may not range over predicates and functions but in higher order logics variables may range over functions and predicates, for example consider this function definition:

MAP2 (f) (x,y) [n] = f(x(n), y(n))

MAP2 is a function which takes as its arguments a function, f, two arrays, x and y, and an implied parameter, n. It will return an array each element of which is the application of the function, f, to the respective elements of x and y.

To be more concrete the function

MAP2 (and) (x,y)

will 'and' together, in turn, the first elements of x and y then the second elements and so to give a new array of boolean values.

The justification for using a higher order logic to describe hardware specification lies in the style of specification which it supports. The use of high order functions leads to compact and comprehensible specifications; it allows time abstractions and the generation of structure. The power of higher order logic is such that most ordinary mathematics, though usually formalised in set theory, can be formalised using it. This is the foundation on which the HOL system is built, Gordon (6).

In HOL there is the idea of a 'Theory'. A theory is created for a specific application and defines the types and constants which are available in that theory. Within a theory there are 'Theorems'. Theorems are either built up, using inference rules, from other theorems or else they are 'axioms', that is they are postulated to be theorems. New 'theories' are developed incrementally from other theories. Thus a theory can have one or more parent theories and all theories are constructed fundamentally from the two standard theories. BOOL, the theory of Booleans, and IND, the theory of individuals.

The HOL system (6) is an interactive proof generator with mechanisms for setting up new theories. In the HOL system we have the essential requirements of abstraction and proof construction needed for digital system verification.

As a concrete example let us return to the latch circuit examined previously.

Given a suitable theory and using the syntax
a → b|c
which is the HOL form of IF a THEN b ELSE c

then the latch considered previously can be described thus:

First define an existence predicate:

prev Xt ≡ ∃t'.(t' ≤ t ∧ Xt')

The Latch behaviour is defined:

LATCH (W,I) OUT ≡
∀ t. (prev Wt ⊃ OUTt = Wt → It|OUT(t − 1))

The more complex latch becomes:

REG (R, W, I) OUT $\equiv$ $\exists$ X. $\forall$ t.
      (prev Rt $\land$ prev Wt) $\supset$
        OUTt = Rt $\rightarrow$ Xt$|$OUT(t $-$ 1)
           $\land$
      Xt = Wt $\rightarrow$ It$|$ X(t $-$ 1)

The description of the implementation structure is:

REGIMP (R, W, I) OUT $\equiv$ $\exists$ X.
      (Latch (W, I) X $\land$ Latch (R, X) OUT)

unfolding the Latch invocations gives:

$\equiv$ $\exists$ X.
      $\forall$ t. (prev Wt $\supset$ Xt = Wt $\rightarrow$ It$|$X(t$-$1))
          $\land$
      $\forall$ t. (prev Rt $\supset$ OUTt = Rt $\rightarrow$ Xt$|$OUT(t$-$1))

Now it can be shown that:

$\exists$x. ($\forall$ t. A $\supset$ B$\land$ $\forall$ t. C $\supset$ D)
          $\Rightarrow$
$\exists$x. ($\forall$t. (A $\land$ C) $\supset$ (B $\land$ D))

so:

REGIMP (R, W, I) OUT $\Rightarrow$ $\exists$ X. $\forall$ t.
      (prev Rt $\land$ prev Wt) $\supset$
        OUTt = Rt $\rightarrow$ Xt$|$OUT(t$-$1)
           $\land$
      Xt = Wt $\rightarrow$ It$|$X(t$-$1)

This 'implies' relationship ($\Rightarrow$) means that the implementation behaviour is more constrained than the specification behaviour but not inconsistent with it. To see what this means let us look first at the definition of Latch. Our formal definition of Latch does not exactly describe the behaviour of the informal description given previously (section 14.3). The informal description suggests that the output is constant, even if its value is not known, before the first write. The formal description does not constrain the Latch output before the first write.

Similarly the REG definition given in this section does not constrain the output until both read and write have occurred. However REGIMP will actually output a constant, but not known, value if the first read precedes the first write.

Alternatively we might be prepared to accept that time ranges over negative value as well as positive and formulate our behaviour descriptions thus:

The Latch behaviour is defined:

LATCH' (W, I) OUT $\equiv$
$\quad \forall t.(OUTt = Wt \rightarrow It|OUT(t-1))$

The more complex latch becomes;

REG'(R, W, I) OUT $\equiv$ $\exists$ X. $\forall$ t.
$\quad$ OUTt = Rt $\rightarrow$ Xt|OUT(t-1)
$\qquad \wedge$
$\quad$ Xt = Wt $\rightarrow$ It|X(t-1)

The description of the implementation structure is:

REGIMP'(R, W, I) OUT $\equiv$ $\exists$ X.
$\quad$ (Latch' (W, I) X $\wedge$ Latch' (R, X,) OUT)

unfolding the Latch invocations gives:

$\equiv$ $\exists$ X.
$\quad \forall$ t. (Xt = Wt $\rightarrow$ It|X(t-1))
$\qquad \wedge$
$\quad \forall$ t. (OUTt = Rt $\rightarrow$ Xt|OUT(t-1))

Now it can be shown that:

$\exists$ x. ($\forall$ t. A $\wedge$ $\forall$ t. B $\Longleftrightarrow$ $\forall$ t. (A $\wedge$ B))

so:

REGIMP' (R, W, I) OUT $\equiv$ $\exists$ X. $\forall$ t.
$\quad$ OUTt = Rt $\rightarrow$ Xt|OUT(t-1)
$\qquad \wedge$
$\quad$ Xt = Wt $\rightarrow$ It|X(t-1)

The equivalence relation ($\Longleftrightarrow$) says that the specification implies the implementation *and* that the implementation implies the specification so the behaviours are the same.

Even though the essence of the HOL system is that theorems once proved are available thereafter the theorem building task is somewhat daunting.

Another approach to specification and verification using higher order logic is the Veritas system, Hanna and Daeche (7). An alternative to the proof approach is the transformation approach. This allows the engineer to manipulate a design in such a way that the semantics of the design remain unchanged but the form is altered. An example is the LTS system.

14.5 LTS

LTS is a language for describing the Behaviour and Layout of VLSI Systems. The LTS system is the language and its associated transformation tools. In LTS signals are entities which vary over time, not fixed values independent of time. Behaviour is the relation of input signals and output signals of this form. In addition, the concept of time is a backward looking model.

Consider the following fragment of LTS:

AtMostRecent (x,y) =
    If x is
        True then y,
        False then AtMostRecent (Last (x), Last (y))

This recursive definition determines the current value of the function AtMostRecent by determining the value of y at the most recent time in the past when the value of x was true. 'Last' is a primitive of the language and 'steps back' one 'chronon' of time. The value of the chronon is in arbitrary units so time is just a discrete ordering with no fixed relation to real time.

This AtMostRecent function could be used to define the elementary latch circuit thus:

Latch (Write, Input) = AtMostRecent (Write, Input)

Using the manipulator we can 'point' at the instance of AtMostRecent here highlighted and perform the unfold transformation. This would transform the description into:

Latch (Write, Input) =
    If Write is
        True then Input,
        False then AtMostRecent (Last (Write), Last (Input))

Using the manipulator to fold the highlighted invocation of AtMostRecent with the definition of latch gives:

Latch (Write, Input) =
    If Write is
        True then Input,
        False then Latch (Last(Write), Last(Input))

Thus we have a recursive definition of Latch. Attempting to implement this function on silicon would require a non-terminating recursion of nested latches. However as LTS is a functional language without side effects applying the primitive delay 'last' to the inputs of a function has the same effect as delaying the output. Thus undistributing both invocations of 'Last' in the above expression gives:

Latch (Write, Input) =
    If Write is
        True then Input,
        False then Last (Latch (Write, Input))

Rearranging to give a name to the output of the function Latch gives:

Latch (Write, Input) =
    If Write is
        True then Input,
        False then Last (L)
    Where L = Latch (Write, Input)

The significance of this change is that in LTS there is a 'standard' interpretation of behaviour descriptions. Under this interpretation the first description could not be laid out but the second can. The block plan of this layout can be seen in Fig. 14.2.

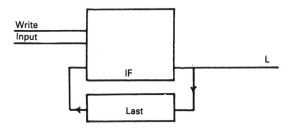

Fig. 14.2   *A latch block plan*

IF is a select function and LAST a chronon delay.

In terms of the complex and simple Latch of Fig. 14.1 the overall function is, assuming definitions of Latch to be available:

Reg (W, R, I) = Y
    Where Y =
        If R is
            True then X,
            False Last (Y)
    Where X =
        If W is
            True then I,
            False then Last (Latch (W, I))

Folding X with Latch and eliminating X gives:

Reg (W, R, I) = Y
    Where Y =
        If R is
            True then Latch (W, I),
            False then Last (Y)

Folding Y with Latch and eliminating Y gives:

Reg (W, R, I) = Latch (R, (Latch (W, I)))

which is the implementation required.

A manipulator for the current LTS system has been implemented. It is a transformation tool which can transform designs into other designs which have the same behaviour. It implements strict equivalence and the related side conditions, of the transformation, are checked automatically. It is embedded in a dialect of the ML language. This allows the user to build up transformation procedures based on the primitives allowed by the manipulator. The language and the system are described in more detail in Babiker et al (8) and (9). The new standard definition of ML is Milner (10).

The limitation of the LTS system at the present is the adherence to strict equivalence. When a design is nearing final implementation, strict equivalence is the right paradigm for exploring alternative structures for implementation efficiency. However, at earlier stages of the design process the engineer may, rightly, be making design trade-offs which limit the possible implementations to a subset of those encompassed by the original description. In such cases it needs to be shown that the proposed changes are within the original specification. This requires conditional equivalence and the ability to verify the conditions. This implies that at least some of the generality of the HOL system must be introduced.

14.6 Conclusions

The formal verification of digital systems can only be satisfactorily achieved if the systems themselves are described in formally based languages. The use of transformation systems can convert easy to understand descriptions into easy to lay out designs. An effective tool which is to encompass a reasonable portion of the design process must allow conditional equivalence.

14.7 Acknowledgements

I am indebted to many people at STL for what I have learnt of formal methods. This is particularly true of the members of the LTS project and I would like to pay a particular tribute to Roger Fleming and Robert Milne.

14.8 References

1   DENVIR, B. T.: 'A lattice-theoretic approach to system specifications and the contractual methodology'. STL Doc NO. 604/stl/tnr 18

2 GORDON, M. J.: 'LCF – LSM', Technical Report No. 41, University of Cambridge Computer Laboratory

3 GORDON, M. J.: 'Proving a computer correct', Technical Report No. 42, University of Cambridge Computer Laboratory

4 GORDON, M. J.: 'Register transfer systems and their behaviour', 1981, 5th Int. Conf. on computer Hardware description languages and their application

5 BARROW, HARRY G.: 'Verify: a program for proving correctness of digital hardware', 1984, *Artificial intelligence*, **24**, 437–491

6 GORDON, M. J.: 'A machine oriented formulation of higher order logic', Technical Report No. 68, University of Cambridge Computer Laboratory

7 HANNA, F. K. and DAECHE, N.: 'Specification and verification using high-order logic', 1985, 7th Int. Conf. on Computer Hardware Description languages and Their Application

8 BABIKER, S. A., FLEMING, R. A., MILNE, R. E.: 'A tutorial for LTS', 1985, STL ITM No 311. 85.4

9 BABIKER, S. A. *et al.*, 'A guide to designing with LTS', 1985, STL ITM No 311.85.5

10 MILNER, R.: 'The standard ML core language (revised)', June 1985, Edinburgh University

Chapter 15

Hardware CAD tools

A. P. Ambler

15.1 Introduction

Hardware designed solely with the objective of performing a particular CAD task, but at a much higher speed than could be accomplished on a conventional computer, is now well established as a concept and a marketable product. Certainly, when one appreciates the excessive and prolific use of computer resources used by, say, logic simulation (see Table 15.1) and other tasks on today's problem sizes, CAD managers must be horrified by the anticipation of the widespread use of VLSI devices and the forthcoming WSI. Unfortunately, leaps and bounds in processing technologies also means leaps and bounds in the required design time, i.e. cpu hours/days. Thus the stage is set for any means to reduce the design bottleneck – the von Neumann machine.

Table 15.1 *CPU times for CAD tasks of differing complexities*

CAD task	Algorithmic complexity	CPU time for equal level of design accomplishment		
		1K Gates	10K Gates	100K Gates
Logic simulation	$0\,(N^2)$	1 Hr.	4 Days	1·2 Yrs
Fault simulation	$0\,(N^2)$	1 Hr.	42 Days	116 Yrs
Maze router	$0\,(N^2)$	20 mins	33 Hrs	138 Days

This chapter will not consider what might seem to be the obvious solution of replacing your home computer with a CRAY, or adding a floating-point co-processor or array processor, but examine hardware that has been designed for the solution of CAD algorithms in particular. Such hardware has the potential benefit of reducing a 42 day logic simulation time to less than 10 minutes!

Logic simulation is an area which has received the most attention from accelerator designers and resulted in the well known examples from Zycad (1) and IBM (2). Thus the major portion of this chapter will concentrate on the acceleration of logic simulation indicating how the potential benefits are achieved.

15.2 Acceleration of logic simulation (3)

Most logic simulation accelerators attempt to exploit two main sources of concurrency: circuit concurrency and algorithm concurrency.

CIRCUIT CONCURRENCY results from the simultaneous processing of the signals which in a real digital circuit would be propagating simultaneously in different parts of the same circuit. Fig. 15.1 shows a possible VLSI device where signals are likely to be changing at the same time in, say, the register and the ALU. This can be simply exploited by a multi-processor machine. In an n-processor machine, the circuit is divided such that each processor separately and concurrently simulates approximately 1/n th of the circuit using standard logic simulation algorithms. The resultant simulator architecture is shown in Fig. 15.2. The partitioning of the circuit in practice may be more complex than a mere geometric 'carve-up , and divisions along functional lines may be more appropriate. Thus, referring to Fig. 15.1, the registers might be assigned to one processor, whilst the ALU is assigned to another. Signals that originate in one part of the circuit being simulated that control or are passed to another separate part of the same circuit being simulated, would be transferred between simulation processors by means of a common bus or inter-processor switch.

ALGORITHM CONCURRENCY is the exploitation of those aspects of the simulation algorithm that are independent and can be processed simultaneously on dissimilar data. The sequential steps involved with the simulation algorithm that might normally be carried out in a software simulator, e.g. accessing the next event from the event queue memory, updating logic values in the logic value memory, and evaluating the new logic element outputs as a result of these input value changes etc., can be broken down even further into a sequence of simple processes and linked together to form a pipeline (4). Thus many activities within a circuit can be processed concurrently, but at different stages within a simulation algorithm, internal to the pipeline. There can be as many concurrent operations as there are pipeline stages, Fig. 15.3.

It is of course possible to have an architecture that exploits both circuit and algorithm concurrency by replacing the independent logic simulation processors in Fig. 15.2 with pipeline processors. Thus it should be possible to take maximum advantage of the simulation process.

To give an indication of the speed of such machines, Table 15.2 lists some published performance figures of a selection of accelerators (2).

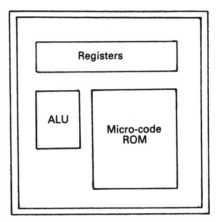

Fig. 15.1   *Integrated circuit to be simulated*

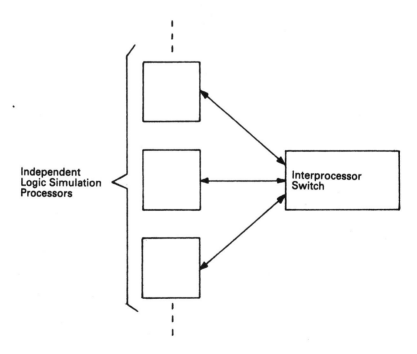

Fig. 15.2   *Multi-processor accelerator architecture*

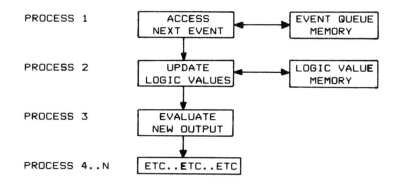

Fig. 15.3   *Pipeline exploiting algorithm concurrency*

The figures quoted compare with a typical 1–5000 gate evaluations per second for a simulator running on a conventional computer.

Table 15.2   *Performance figures for selected accelerators*

Accelerator	Gate evaluations/sec
IBM	960 Million
ZYCAD	3 *Billion*
VALID	0·5 Million
DAISY	0·1 Million

It is, perhaps, worth mentioning at this point the difference between the different units that are sometimes quoted in reference to the performance of these accelerators. Refer to Fig. 15.4. When a logic value change occurs on the output of gate A, this is referred to as an 'event'. Thus the outputs of these gates must be evaluated – gate evaluation. Once all three gates have been processed, an event evaluation has been completed, i.e. a gate evaluation processes just one gate, whereas an event evaluation processes all gates on the fanout of a logic gate change. Thus, 100 event evaluations per second can be reckoned to be faster than 100 gate evaluations per second.

In order to achieve the maximum speed improvements, the CAD algorithms are hardwired – no portion of the algorithm is contained in software, if at all possible. These machines are generally referred to as 'point' accelerators. Unfortunately, such 'permanence' in the algorithm can be a disadvantage. It might reasonably be expected that the CAD algorithms being used will be updated and refined at not infrequent intervals, e.g. if a software-based simulator had been purchased with a maintenance contract. To obtain a machine with no easy way to update these algorithms might then be considered to be an expensive luxury, or so the protagonists of general purpose accelerators would have you believe!

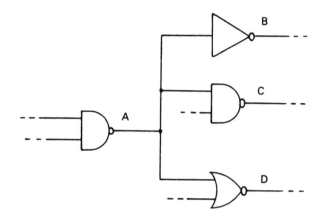

Fig. 15.4   *Comparison between 'event' and 'gate' evaluation*

15.3 Bottlenecks

The potential benefits to be accrued from the use of such machines is obvious. However, their disadvantages must also be highlighted.

The performance gains may not be as great as a potential purchaser of such a system might have anticipated. The quoted speed-ups usually relate to the processing power of the accelerator hardware in isolation (see Fig. 15.5), leaving out transmission of simulation data to and from the accelerator and the host computer, i.e. the i/o bandwidth. These tasks form part of the overall simulation process and their times of execution must be included in any

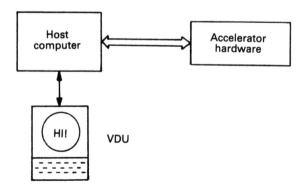

Fig. 15.5   *Typical accelerator configuration*

benchmark. Unfortunately, they can often dominate the overall simulation time and indeed, can create such a bottleneck that the performance gains can be negative, Fig. 15.6.

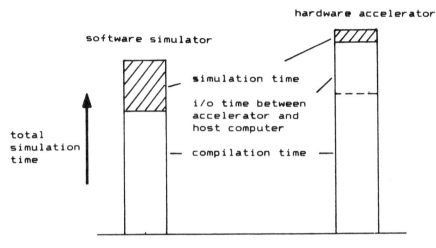

The advantages of using an accelerator can be dependent upon the size of problem applied to it. The added compilation overheads and input/output time can mean that the use of an accelerator can be detrimental even though the raw simulation process time has been reduced. It is only if the raw simulation time is significant in comparison with these other factors that an accelerator becomes useful.

Fig. 15.6   *Potential bottlenecks in the use of accelerators*

IBM with their own in-house developed machine have had reported the following figures (2):

Consider the simulation of a 500K gate processor over a 100 instruction sequence. Using the standard software, the simulator required 4·5 minutes, whilst the hardware accelerator required 49 minutes! Repeating the simulation but over a 1 million instruction sequence required 250 hours on the software simulator, but only 66 minutes on the accelerator.

Other major users of accelerators have reported similar problems (5).

The usefulness of these types of accelerators can therefore be seen to be applicable only to problems of above a certain size.

The use of accelerators can also raise attention to other overheads that were not immediately apparent before. Bic Wood (6) points out other performance figures associated with the use of an accelerator: an accelerator attached to a workstation performed a simulation in just 11 minutes whilst the compilation, linking and flattening performed on the workstation to prepare the simulation data for the accelerator took 13 hours!

Whereas compilation, linking, flattening and simulation without the accelerator would have required perhaps days of cpu time, the focus of attention is now on the 'dead time' associated with the workstation intensive tasks. The requirement will now be to accelerate these computational aspects of the process.

15.4 General purpose accelerators

Fig. 15.7 shows a possible scenario with, admittedly, an arbitrary selection of CAD processes and durations representing the total design time for an integrated circuit.

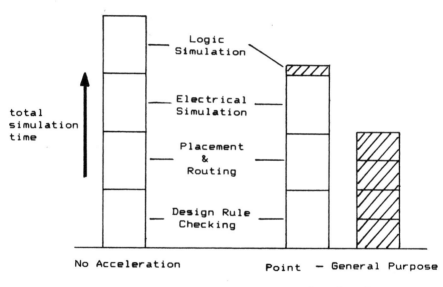

The above shows a scenario with, admittedly, an arbitrary selection of CAD processes and durations representing the total design time for an integrated circuit. Comparison is made between no acceleration, use of a point accelerator, and general purpose accelerator. Obviously, if the proportion of time that logic simulation represents is significant, then a point accelerator can be justified. General purpose accelerators promise moderate reduction in processing times for most CAD tasks. Choice of accelerator needs to be based on a detailed 'economic' comparison.

Fig. 15.7   *The case for general purpose accelerators*

This shows a possible argument against the use of point accelerators, i.e. that the complete design process for any integrated circuit consists of more than just logic simulation. Depending upon the relative proportion of logic simulation compared with all the other design processes, the possibly large outlay in such a machine may not seem economic to some users.

There are proposed and existing designs for the acceleration of other CAD algorithms, this subject is not limited to logic simulation. The list of algorithms that are being catered for in special purpose hardware is extending every day, but does include: logic simulation, fault simulation, test pattern generation (3) (7), electrical circuit simulation (8) (9), placement and routing (10), design rule checking (11), to name but a few. The inference here being that to cover the whole design process would require the purchase of a number of machines.

However, if a machine were available that could accelerate several, if not all, the CAD algorithms, then a more total solution might be found. Such a machine may not need to accelerate at anything like the possible achievements of mainframe point accelerators in order to compete. Fig. 15.7 demonstrates this point for a general purpose accelerator which speeds-up all CAD tasks by only a factor of two. In order to accelerate many algorithms, it is obvious that rather than being hardwired, the algorithms are kept in software. At the time of writing, several such machines are being announced. These machines are effectively faster single processors optimised for particular software code.

Such machines do, though, have a long way to go before they can begin to approach the performances of the point accelerators.

Several workers are known to be approaching this problem by applying multiple processors to the solution (3) (8). Here, though, other problems can arise as indicated by Dally (12). In this work, the design of an accelerator for switch level simulation using multiple processors utilised a common bus for intercommunication between the processors. A mathematical model was developed to establish the likely operating parameters of the machine. The results quite clearly showed that bus contention was a limit to the performance and that the optimum number of processors for a given problem size would reach a peak and then decline as the size of the problem increased.

This is hardly a satisfactory state of affairs when the requirement is to be able to apply more processing power to the larger problems – an obvious problem which has to be tackled.

15.5 Selection of accelerator

Selection of the particular machine for your own application is extremely difficult, given the large number of vendors. Each one will have its merits that must be isolated for comparison purposes. Obviously, if you are already using a particular type of workstation for which there is also a directly compatible accelerator, then that would be the first machine to consider.

Several software-based simulator vendors now market hardware refined to their particular algorithms. This must also be a consideration as the alternative would be to use a third party accelerator with different algorithms producing slightly different results to those previously obtained. This is in much the same way as using Software Simulator A and converting to Software Simulator B which will have different output.

Another point for consideration is whether or not alternative software CAD algorithms exist that will equal or maybe surpass what might be available in hardware. Some algorithms are more susceptible to this than others. Electrical circuit simulation is, perhaps, one such example where use of other algorithms can, although more limiting in their application, achieve respectable acceleration.

However, the baseline for the final decision must be related to price and performance. Performance can only be confirmed after several benchmarks have been applied to the machine(s) being considered.

The selection of the benchmark circuits must be done carefully, preferably after as much analysis of the machine architecture as is possible has been performed. For example, an architecture that consists only of, say, a pipeline, will optimally accelerate circuits that consist of a string of logic gates connected in series with the input node toggling 1-0-1-0..., whereas an n-processor machine which has the circuit to be simulated partitioned between them will optimally accelerate circuits with n active circuit blocks. Obviously, if all your classes of circuit designs for simulation fall into either of these two classes, then the decision is relatively simple. But, in the general case it is not quite so clear-cut, therefore several carefully chosen benchmarks must be tried.

15.6 Conclusions

The use of hardware to accelerate CAD algorithms is an obvious requirement, but as all new forms of hardware costs money, the question must be asked if they are really necessary for all design environments. Often there will be better solutions that can be obtained through less costly means.

For example, many will find that sufficient increase in throughput can be achieved simply by increasing the memory in their existing computer. If they are currently experiencing large amounts of page swapping, the extra memory will remove this bottleneck producing an effective speed-up that could be sufficient for their immediate needs.

In other situations, a change of design strategy might remove the need for an acceleration or speed-up. Consider the designer wanting to put built-in self test into a design using established BILBO methods, i.e. random test pattern generation and signature analysis. The standard approach would be to perform extremely expensive fault simulation to verify the fault coverage of the random test patterns, and another simulation to calculate the resulting fault free signature. This would suggest the need for an accelerator. However, if the random test pattern generator were to be adapted to provide exhaustive tests, the need for fault simulation is now removed. Resorting to 'gold chip' methods to observe the fault free signature removes the need for signature calculation, provided that the quality of test result is acceptable.

Hardware accelerators *are* becoming a necessary and useful aid to the design process – the larger complexities of devices now being designed need vast amounts of computational power to produce the results required which only these new machines can offer. With the alternative being continued reliance upon traditional Von-Neumann mainframe computers, the possibilities are that short-cuts will be made thereby reducing the quality of the product. In the short term, accelerators can offer huge speed-ups of traditional algorithms enabling the design engineers to continue with the use of their familiar tools. Thus a relatively obvious, but 'brute force and ignorance' approach has been adopted.

The longer term suggests that more thought is needed in order to be able to design wafer-scale circuits in a reasonable time. New algorithms must be developed, e.g. formal verification that promises to make simulation redundant.

However, it will always be the case that designers would like the questions they put to their CAD tools to be answered in a shorter time than they are currently getting. Accelerators, then, will be required to speed-up these new algorithms. . . .

References

1 ZYCAD: 'The system development engine – SDE', 1985, product specification brochure

2 BLANK, T.: 'A survey of hardware accelerators used in computer-aided design', *IEEE Design and Test of Computers*, 1984, **1**, 3, 21–39

3 AMBLER, A. P., MANNING, R. L. and MUHAMMED, N.: 'Hardware accelerators for CAD', *IEE Computer-Aided Engineering Journal*, 1985, **2**, 3, 110–117

4 GLAZIER, M. E. and AMBLER, A. P.: 'Ultimate: a hardware logic simulation engine', Proceedings of ACM-IEEE 21st Design Automation Conference, 1984

5 SMITH, L. T. and REZAC, R. R.: 'Methodology for and results from the use of a hardware logic simulation engine for fault simulation', Proceedings of IEEE International Test Conference, 1984

6 WOOD, B.: 'The A-Series accelerator: a case history', *VLSI Systems Design*, 1985, **6**, 3, 58–59

7 KRAMER, G. A.: 'Brute force and complexity management: two approaches to digital test generation', 1984, M.Sc. Thesis, MIT

8 DEUTSCH, J. T. and NEWTON, A. R.: 'A multiprocessor implementation of relaxation-based electrical circuit simulation', Proceedings of 21st ACM-IEEE Design Automation Conference, 1984

9 MANNING, R. L. and AMBLER, A. P.: 'High-speed simulation of 1 million transistor circuits', Proceedings of IEEE Custom Integrated Circuits Conference, 1985

10 LOO, C. K. P.: 'Integrated circuit array processor for a hardware router', Ph.D. Thesis, University of Manchester, 1984

11 SEILER, L. D.: 'A hardware assisted methodology for VLSI Design rule checking', Ph.D. Thesis, MIT, 1985

12 DALLY, W. J.: 'The MOSSIM Simulation engine: architecture and design', California Institute of Technology, 1984

Index

A

ADA, 116
  package, 117
ALU, see Arithmetic Logic Unit
APECS, 113
  communication, 113
ASM, see Algorithmic State Machine
ATE, 63
Abstract data types, 117
Abutment, 67
Accelerators, 178
  algorithm concurrency, 179
  alternatives to, 185
  benchmarks, 186
  bottlenecks, 182
  circuit concurrency, 179
  general purpose, 184
  logic event/gate evaluation, 181
  logic simulation, 179
  multi-processor, 179, 185
  performance, 181
  pipeline, 179
  point accelerators, 181
  selection of, 185
Algorithmic State Machines, 129
Algorithms for abutment, 92
Algorithms for spacing, 88
Application-specific ICs, 138
Applicative languages, 120
Arithmetic Logic Unit, 126
Assertions, 117, 121
Associative Memory, 127
Automated design techniques, 95

B

BILBO, 132
BIST, 50
BLAM, 130
Back-annotation, 101

Backward differentiation formula, 21
Behavioural model, 40
Bilateral transmission gate, 40
Bipartite Folding, 130
Bit Map, 128
Block, APECS, 113
  STRICT, 120
Bristle block, 108
  Cell, 109

C

CAD concepts, 11
  system, 148, 160
    system bottom-up, 151
    system top-down, 151
CIF, 107
CMOS, 128
Cells, 107, 109, 113
  CIF, 109
  boundary problems, 103
  external description, 104
  full instantiation, 103
  overlapping, 103
  re-mapping, 103
  standard, 96
Charge Sharing, 128
Chip assembly, 90
  methods, 91
Circuit comparison, 100
Circuit verification, 96
Cluster development, 68
Code word, 133
Compactor, 105
  early, 88
  graph based method, 89
Companion model, 22
Compile code simulator, 33
Complexity, 135, 138
  management strategy, 163
  pyramid representation, 138

Component, 112
  APECS, 113
Computer Aided Test (CAT), 45
Computer execution time, 102
Conflicts, 76
Conformance, 167, 168
Congestion, 68
Content-addressable Memory, 127
Contraction mapping, 22, 30
Contractual Model, 167, 168
Controllability, 132
Correctness by construction, 136, 137, 143
Cost estimation, 68
Crosspoint fault, 132
Cube notation, 128

D

D-algorithm, 57
  D-drive process, 57
  D-intersection, 57
  PDC, 57
  PDCF, 57
  consistency operation, 57
DA Programming, 15
DA or CAD, 12
DA systems, 15
DA tools, 15
  design capture, 15
  high level specification languages, 15
  layout tools, 16
    simulation, 16
    static validation, 16
    test generation, 16
DA versus CAD, 11, 12
Data management system, 161
Data model, 155
  canonical schema, 157
  third normal form, 156, 158
Data objects, 156
  attribute, 156, 158
  entity, 156, 158
Database, circuit, 101
  interface programs, 159
  layout, 104
Database management system, 151
  data definition, 152
  features, 152
  libraries, 152
  protection, 152
Database schema, 155
  subschema, 155
  symbolic, 104
  unified, 104, 105

Decoder, 126
Degating of PLA Inputs, 126 133
Depth first, 73
Design activities, 148
  areas, 11
  control, 163
Design data, 153
  behavioural, 153
  physical, 154
  stuctural, 154
Design for testability, 45, 46, 139
  ad-hoc, 46
  neo-structured methods, 48
  scan path, 47
  structured methods, 47
  techniques, 46
Design lag, 10
  leverage, 138
  rules, 97
    electrical, 99
Design tasks, 10, 11, 14
Design unit, 163
  network, 164
  outline, 164
  symbol, 164
Designer restrictions, 139
Device recognition, 99
Differential equations, 19, 20, 28
Domino Logic, 128
Dynamic PLA, 128
  state, 39

E

ESPRESSO-11 logic minimiser, 146
Euler backward integration formula, 21
Evaluation of logic changes, 35
Event scheduling techniques, 36
Examples, SRCELL, 107
  Srcell, 114
  Srpair, 121
  adder – resource, 117
  adder 4b, 118
  equivalence, 115
  equivalence model, 116
  flip-flop, 119, 120
  fourbitadder, 117, 118
  inner, 119
  purestructure, 118
  shiftreg, 111
Exclusive-or, 126
Expert systems in silicon compilation, 145

Extraction, capacitance, 97, 101
  circuit, 99
  parameter, 101
  resistance, 97, 101
  wire delay, 96

F

FIRST, 139, 140, 145
  interface, 139
  Silicon Compiler, 139
Fault coverage, 46
Fault simulation, 52
  concurrent, 53, 54
  deductive, 53
  parallel, 53
Faults, pattern sensitive, 50, 52
Faults, stuck-open, 50, 51
Feedback in silicon compilation, 146
File-based systems, 161
  advantages, 161
  disadvantages, 161
Finite State machine, 126
Floor plans, 85
  high level design, 87
Folding, 128, 130, 131
Formal methods, 167
  verification, 167
Format, logical, 109
  structural, 109
  temporal, 109
Full custom, 97
  Design, 124
Functional, form, 112
  modelling language, 40
  structure, 113
Future DA tools, 16
  silicon compilers, 16
  AI techniques, 17

G

GAELIC, 107
Gate Arrays, 67, 96, 124
Gate delays, 37
  ambiguity, 38
  inertial, 37
  load dependent, 38
  rise and fall, 37
  transport, 37
Gate recognition, 99
Gaussian elimination, 24
Gauss-Jacobi algorithm, 26, 27, 29
Gauss-Seidel algorithm, 26, 27, 29
Generator Development Tools (GDT), 143

Genesil, 144
Graph-theoretic floor-planning, 145
Graphics editor, 105

H

HOL – Higher Order Logic, 170
  'Theorems', 171
  'Theory', 171
Hardware Description Languages (HDL),
    116, 136
  CLASP, 145
  FIRST, 142
  L, 143
  Model, 142
Heuristic Method, 130
Heuristic algorithm, 102, 130
Heuristics, 74
Hierarchical partitioning, 25
Hierarchy, 102
  corresponding, 103
  layout verification, 102
High level description, 108, 109

I

IC design system evolution, 150
IC design system requirements, 151
ILAP, 110
Integrated database system, 162
  advantages, 162
  disadvantages, 162
Intermediate Format, 107
Iteration, 130

J

Jacobian matrix, 22, 24

L

LAP, 110
LSM – Logic of Sequential Machines, 168
  Behaviour expression, 168
  Composition theorem, 170
LTS – Language for behaviour and layout of
    hardware, 173
  Manipulator, 174
Laplace's equation, 101
Latency, 25
Layout expansion, 78
Lee's algorithm, 72
Left edge algorithm, 75
Levels of design, 14
Logic Minimisation, 129

M

MINI, 130
MODEL, 119
    procedural abstraction, 119
MOSIS, 138, 143
Macro Cell, 129
Mincut algorithm, 70
Modelling analogue functions, 41
Modula-2, 113
Module, model compiler, 33
    preprocessor, 33

N

Newton-Raphson algorithm, 21, 22
Next event list processing, 36
Nodal equations, 19
Non-orthogonal shapes, 102

O

Observability, 132
Off line self-test, 49

P

PART, 132
PLA, see Programmable Logic Array
PLA/TG, 131, 132
PLAFOLD, 130
PLAP, 110
PLEASURE, 131
PODEM-X, 59
    FFSIM, 59
    PODEM, 58
    RAPS, 59
    SRTG, 59
Parasitic values, 101
Path sensitisation, 56
Placement and route, 67
Placement improvement, 70
Placement relative, 77
Programmable Logic Array, 124
Pseudo-NMOS, 128

R

Random Test Patterns, 132
Regular Structures, 125
Relational database, 155
Relaxation methods, 19, 25–30
Residue Code, 133
Rightness, 167

Routers, Channel, 75
    Global, 76
    Hierarchical, 76
Rule checking, dimension, 97, 102, 103
    electrical, 99
    inter-cell, 103, 105
    intra-cell, 103
    real-time, 105

S

SLA, see Storage/Logic Array
SLIC silicon compilation tool, 144
STRICT, 120
    assertions, 121
    behaviour, 120
    blocks, 120
    generic parameter, 121
    guards, 121
    pragmats, 121
    structure, 121
    type definition, 121
Scan Path, 132
Scan design problems, 49
Schichman integration formula, 21
Selective trace simulation, 35
Semi-custom, 96
Shape operations, 98
Signature Analysis, 132
Silicon compilation, 95, 108, 136–138
Silicon compilers, 82, 139–144
    CAPRI, 145
    Chipsmith, 140, 142
    Concorde, 144
    FIRST, 139, 140
    Genesil, 144
    MacPitts, 142
    MetaSyn, 142
    PLEX, 143
    SILI, 145
    behavioural input, 145
    bristle blocks, 145, 108
    comparison with software compilers, 136,
        137
    expert systems, 145
    feedback, 146
    floorplan, 137, 145
    influence of Carver Mead, 144
    structural input, 145
    tools, 143
    use of hierarchy, 137
    structures project, 140
Simulated annealing, 70

Simulation, 32, 129
  algorithm, 35
  anomalies, 39
  backward, 56, 57
  modelling, 33
  process, 33
  uses, 32
Simulators block level, 32
  functional level, 32
  gate level, 32
Special purpose CAD hardware, 17
Standard Cell, 124
Static PLA, 128
Storage/Logic Array, 127
Symbolic cell design, 87
  design, 81, 82
    objective, 85
Symbolic methods, 82
  coarse grid, 82
  gate matrix, 83
  sticks, 83

T

TPLA, 132
Table driven simulator, 33
Test Generation, 131
  techniques, 55, 59
    algebraic, 55
    structural, 56
Test problem, 45
Testability, 46, 132
  HI-TAP/Camelot, 62
  SCOAP, 47, 60
  aids, 59
  analysis, 46, 60
  controllability, 46, 60
  observability, 46, 60
  rule checking, 62

Testing costs, 44
  reduction, 44
Time mapping algorithm, 36
Tools for silicon compilation, 143, 144
  GDT, 143
  Genesil, 144
  SLIC, 144
Transformation Tools, 168, 173, 176
Transient analysis, 19
Truncation error, 21 22

V

VHDL, 116
  Architectural body, 118
  assignment, 118
  behavioural body, 117, 118
  control flow, 117
  data flow, 118
  loops, 118
  selection, 118
  signal, 118
VHSIC, 116
VLSI CAD, skills, 13
  data, 14
  history, 13
  trends, 10
Verification, 129
Veritas, 173
Virtual grid, 89, 105

W

Waveform relaxation, 28–30
Weinberger Array, 143
Weinberger PLA, 126
Wired logic, 39
Wiring channels, 67

Y

Yield, 46